Advances in Biosensors: Reviews

Volume 1

S. Yurish
Editor

Advances in Biosensors: Reviews

Volume 1

International Frequency Sensor Association Publishing

S. Yurish, *Editor*
Advances in Biosensors: Reviews. Volume 1.

Published by IFSA Publishing, S. L., 2017
E-mail (for print book orders and customer service enquires):
ifsa.books@sensorsportal.com

Visit our Home Page on http://www.sensorsportal.com

ISBN: 978-84-697-3467-4
e-ISBN: 978-84-697-3468-1
BN-20170531-XX
BIC: TCBS

Contents

Preface

According to market research the global sensors market was valued at $113.2 billion in 2016, and should reach $190.6 billion by 2021, a compound annual growth rate (CAGR) of 11 % over the five-year period. Biosensors market is expected to be valued at $27.06 billion by 2022, growing at a CAGR of 8.84 % between 2017 and 2022. The market growth is driven by the continuous technological advancements in the biosensors ecosystem, increase in the use of biosensors for nonmedical applications, lucrative growth in point-of-care diagnostics (cardiac markers, infectious diseases, coagulation monitoring, pregnancy and fertility testing, blood gas and electrolytes, tumor or cancers markers, urinalysis testing, and cholesterol tests, and others), and rise in the demand for glucose monitoring systems. But it is no any doubts that these advancements are based on long-term researches and developments.

Every research and development in biosensors (as well as in any other research fields) is started from a state-of-the-art review. Such review is one of the most labor- and time-consuming parts of research, especially in high technological areas as biosensors and chemical sensors. It is strongly necessary to take into account and reflect in the review the current stage of development, including existing sensing principles, methods of measurements, technologies and existing devices. Many PhD students and researchers working in the same area must make (and do it) the same type of work. A researcher must find appropriate references, to read it and make a critical analysis to determine what was done well before and what was not solved till now, and determine and formulate his future scientific aim and objectives.

To help researchers save time and taxpayers money, we have started to publish 'Advances in Sensors: Reviews' Book Series. Since 2012, this Book Series has become very popular among sensor community, and volumes 5 and 6 are now in our editorial portfolio. Taking into account the importance of this Book Series, since 2016 it is published as open access books. In 2017, we have decided to start publication of the second open access Book Series titled 'Advances in Biosensors: Reviews'. I hope it will be also very useful as the first Book Series.

The first volume of 'Advances in Biosensors: Reviews', Book Series contains 7 chapters written by 14 authors from 9 countries: Australia, Bulgaria, China, Germany, Poland, Russia, Spain, Turkey and USA.

Chapter 1 describes printed and flexible bio- and chemical sensors. Printed electronics, in general, and printed electronic sensors in particular is an ever expanding field. The capability of using additive manufacturing tools and techniques in conjunction with novel sensor designs, such as the SAW biochemical sensors described in this Chapter, leads to a wide range of possibilities in terms of low cost, easy to deploy devices.

Chapter 2 discuss advances in development and applications of up conversion nanoparticle in biosensing. Upconversion nanoparticles (UCNPs) hold great promise in broad applications of biomedicine because of their unique properties of being able to convert near-infrared (NIR) light into luminescence of shorter wavelength (UV, visible or NIR). In particular, this intriguing feature of UCNPs find great potential applications in highly sensitive biodetection by using near-infrared excitation to minimize the background signal and enhance the overall signal-to-noise ratio. Other merits, such as narrow multicolor emission, exceptional photostability, and low cytotoxicity, have also confer UCNP the key advantages in multiplexing and long-term detection in biological settings. In this chapter, the development and applications of UCNP-based biosensors are summarized. A short introduction of UCNPs, including their photoluminescence mechanisms and unique optical properties, is presented in the first part. We then describe the design scheme of UCNP-based sensors, covering the synthesis methods of UCNPs with high emission efficiency, and approaches of coupling the nanoparticles with detection reagents. State-of-art examples of using UCNPs in sensing of ions, biomolecules, and gas molecules are then provided. Lastly, the challenges and future opportunities of the UCNP nanosensors are discussed.

Chapter 3 reviews nano-particle analysis with the plasmon assisted microscopy of nano-objects sensor (PAMONO-sensor). In this review authors are focused on the hardware and software features of a sensor that applies the surface plasmon resonance (SPR) highly sensitive optical method. In Section 3.2 a short overview of state-of-the-art technical aspects of the PAMONO-sensor is given. Subsequently, the ability to detect signals of individual viral or non-biological particles and to determine the concentration of particles in analyzed samples are discussed. In Section 3.3 authors are focused on the aspects that are highlighted within the analysis-architecture overview. This consists of the sensor data acquisition, feature extraction, classification and optimization of the whole pipeline to match specific characteristics of the PAMONO-sensor and the sensor data. The Section 3.4 is outlined selected results of the different approaches. Finally, the results and describe future work are summarized in section 3.5.

Chapter 4 reviews MEMS magnetic sensors with sufficiently high sensitivity, which are able to detect the bio-magnetic fields produced by the biologic tissues or organs, thus providing a non-invasive mean to detect the activity of the living systems.

Chapter 5 describes an application of surface photo-charge effect for control of fluids. The effect has the great potential to complement the already known and used methods for control of fluids with a rapid and contactless method. It provides a universal testing method since a signal is generated by all types of fluids, and the method could be used for characterization of any fluid. The main advantages of the methods are highlighted in this chapter.

Chapter 6 summarizes physiological properties of human nose to be able to see the similarities with an electronic nose. Than the main parts and the features of electronic nose system are explained. Sensors are the essential building blocks of the system that must be selected according to the application area and the chemical gases to be sensed. Also, application areas are introduced and related sensors of some are mentioned. This chapter can be considered as a basis for the Chapter 7.

Chapter 7 reports the achievements on the field of artificial senses, such as electronic nose and electronic tongue. It examines multivariate data processing methods and demonstrates a promising potential for rapid routine analysis. Main attention is focused on detailed description of sensor used, construction and principle of operation of these systems. A brief review about the progress in the field of artificial senses and future trends in concerned. A special attention has been paid to the application of these systems in two dominant fields, namely in food investigations and environmental monitoring.

We hope that readers enjoy this book and that can be a valuable tool for those who are involved in research and development different biosensors and biosensing systems.

Sergey Y. Yurish,
Editor, IFSA Publishing   *Barcelona, Spain*

Contributors

Ozge Cihanbegendi Sahin, Dokuz Eylul University, Department of Electrical and Electronics Engineering, Izmir/TURKEY

Tomasz Dymerski, Department of Analytical Chemistry, Faculty of Chemistry, Gdańsk University of Technology, Gdańsk, Poland

Peter Fuhr, Distinguished Scientist, Oak Ridge National Laboratory, Oak Ridge, TN 37831 USA.

Roland Hergenröder, ISAS, Leibniz Institut für Analytische Wissenschaften, Dortmund, Germany

M. K. Kuneva, Institute of Solid State Physics – BAS, 72 Tzarigradsko Chaussee, 1784 Sofia

Jan Eric Lenssen, Department of Computer Science VII, Dortmund University of Technology, Dortmund, Germany

Liuen Liang, ARC Centre of Excellence for Nanoscale BioPhotonics, Department of Physics and Astronomy, Macquarie University, Sydney, NSW, 2109, Australia

Pascal Libuschewski, Department of Computer Science VII, Dortmund University of Technology, Dortmund, Germany

Marissa E. Morales-Rodriguez, Research Scientist, Oak Ridge National Laboratory, Oak Ridge, TN 37831 USA.

J. L. Pérez-Díaz, Universidad Carlos III de Madrid, Instituto Pedro Juan de Lastanosa, Av. Universidad 30, 28911 Leganés, Madrid

Victoria Shpacovitch, ISAS, Leibniz Institut für Analytische Wissenschaften, Dortmund, Germany

Dominic Siedhoff, Department of Computer Science VII, Dortmund University of Technology, Dortmund, Germany

Fei Wang, Department of Electrical and Electronic Engineering, Southern University of Science and Technology, Shenzhen, 518055, China

Frank Weichert, Department of Computer Science VII, Dortmund University of Technology, Dortmund, Germany

Zhen Yang, School of Physics and Electronic Engineering, Xinyang Normal University, Xinyang, 464000, China

Run Zhang, School of Chemical Engineering, University of Science and Technology Liaoning, 185 Qianshan Zhong Road, Anshan 114044, P. R. China; Australian Institute for Bioengineering and Nanotechnology, The University of Queensland, St Lucia, QLD, 4072, Australia

Andrei V. Zvyagin, ARC Centre of Excellence for Nanoscale BioPhotonics, Department of Physics and Astronomy, Macquarie University, Sydney, NSW, 2109, Australia; Laboratory of Optical Theranostics, Nizhny Novgorod State University, Nizhny Novgorod 603950, Russia

Chapter 1

Printed and Flexible BioChemical Sensors/Electronics

Marissa E. Morales-Rodriguez and Peter Fuhr

1.1. Electronic Sensors

Sensors. While there has been extensive literature published presenting sensor technology fundamentals and a wide array of parameters being "sensed", considerable ambiguity still exists regarding the definition and classification of a "sensor". The traditional approach used in describing a sensor is to either list the properties of the sensed parameter or the technology used in the detection and measurement of the sensed parameter. While the terms "sensor" and "transducer" are frequently interchanged, the American National Standards Institute (ANSI) standard MC6.1 defines a transducer as "a device which provides a usable output in response to a specific measurand". An output is defined as an "electrical quantity," and a measurand is "a physical quantity, property, or condition which is measured." Restated, a sensor or transducer takes one form of energy and converts it into another form. In most instances, the output is an electrical signal. Nuances related to having the sensor/transducer directly performing the energy conversion or having the sensor/transducer use one form of energy to convert another form of energy into yet a third form of energy are best expressed diagrammatically as in Fig. 1.1.

An example of a sensor (or transducer) that converts a mechanical pressure into an electrical signal using a piezoelectric material is depicted in Fig. 1.2.

Marissa E. Morales-Rodriguez
Oak Ridge National Laboratory, Oak Ridge, TN 37831, USA

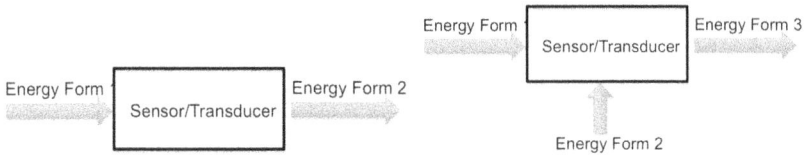

Fig. 1.1. Graphical "definition" of a sensor or transducer.

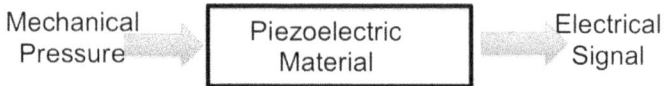

Fig. 1.2. A piezoelectrical transducer (PZT) based pressure sensor.

Frequently a sensor relies on a multistep – or multiple energy transformation – design where the parameter of interest is converted from one form of energy into another which in turn is converted into yet another form of energy. An example of such a sensor "system" is a fiberoptic magnetic-field sensor in which a magnetostrictive jacket (wrapped around an optical fiber) is used to convert a magnetic field into an induced strain within the optical fiber. The resulting change in the fiber's gauge length is measured interferometrically with an end-stage optical-to-electrical detector. The end result of this multistep transduction process is an electrical signal whose signal level varies in (some) accordance with the magnetic field strength. A representation of such a multistep sensor is presented as Fig. 1.3.

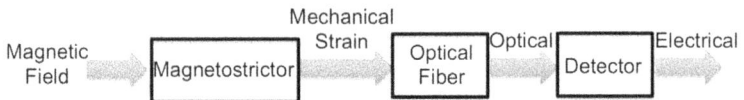

Fig. 1.3. A magnetic field sensor with an electrical output signal.

The illustrative examples of sensors presented in Figs. 1.1 – 1.3 mask the reality of an operating device. In the vast majority of instances, the commonly used phrase "sensors" refers to a system comprised of a variable number of components which can be roughly categorized as: sensor element, sensor packaging and connections, and the sensor signal processing hardware and software. These three basic components may, in themselves, be comprised of various levels of sophistication and complexity but combined are described as the "sensor".

The sensing of chemicals and/or biochemicals follows these same principles. A composite view of different types of chemical sensors is presented as Fig. 1.4. Note that while this diagram is organized via a classification of sensing objects, the "energy form 2" is an electrical signal (manifested as a variation in an electrical parameter (such as voltage, resistance, capacitance, etc.).

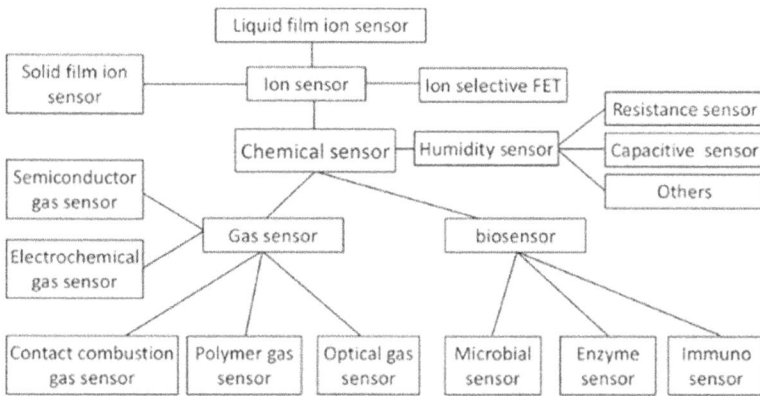

Fig. 1.4. Chemical sensing may take multiple designs, but still align with the classic definition shown in Fig. 1.1.

Fig. 1.4 groups sensors into logical elements (such as "solid film ion sensor", semiconductor gas sensor", etc.) combining to generate the aforementioned variation in an electrical signal (or electrical parameter). The advancements in semiconductor processing in the 1970's onward led to a significant change in the ease with which chemical sensing could be conducted. Furthermore, discrete chemical sensors began being bundled into small assay "kits" culminating in semiconductor wafers with multiple chemical sensors being mass produced. Integration of such devices into complete sensing systems with companion software applications leads to the categorization of such systems as "electronic sensors".

Advancements in printing of electronic structures is resulting in a (potentially) significant change in biochemical sensing.

1.2. Intersecting Technologies and Overlapping Needs

The need for increased availability of sensors to measure and monitor a wide variety of parameters has been widely reported in application areas ranging from environment tracking to industrial sites to local-, regional-, national-, and international-scale monitoring of climate variations. The parameters of interest – that should be measured - outlined in the 2006 Stern Review: The Economics of Climate Change, are strikingly similar to those measurement parameters associated with industrial operations: chemicals, biologicals, and classic physical parameters (temperature, pressures, etc.).

Therefore given the overlapping sensing needs, technological solutions developed for operation in one setting (or application) may be well suited for easy adoption into an entirely different setting. In all cases, the use of communications and networking technologies allow for measurements to be obtained and catalogued via remote access, thereby removing the cost and time associated with fieldwork sampling. Examples of overlapping needs for biological, chemical and environmental sensors are presented.

1.2.1. An Optical Probe for Biogeochemical Sensing of Peat Bogs

The models for the environmental and climatological "operation" of our planet are typified by the Bretherton earth system model shown in Fig. 1.5. Within this model lies the impacts associated with peat bogs. While the exact details of how these areas impact the entire cycle are being studied by a wide array of researchers, it is well known that peat bogs store and release large amounts of CO_2 and other greenhouse gases each year. Collectively peat bogs are a large source of atmospheric CO_2 and researchers are now beginning to study them to understand their significance in the global carbon cycle. One example of addressing devices to assist in the investigation of biogeochemical processes that result in greenhouse gas emissions from peat bogs involves a novel multi-measurement probe that enables field experiment manipulations to uncover linkages between biogeochemical cycling within ecosystems, controlled laboratory experiments, and mechanistic process representations within models. The goal is to understand better the interplay among soil physical and chemical characteristics, microbial activity and biogeochemical processes in response to varying environmental conditions. Simultaneously monitoring multiple factors affecting carbon cycle processes enables experiments that uncover key mechanisms for model development and validation. The use of a multifunction probe supplies unprecedented measurement and sampling capabilities within the soil profile of intact ecosystems and laboratory mesocosms, providing data that significantly improves understanding on spatial and temporal scales.

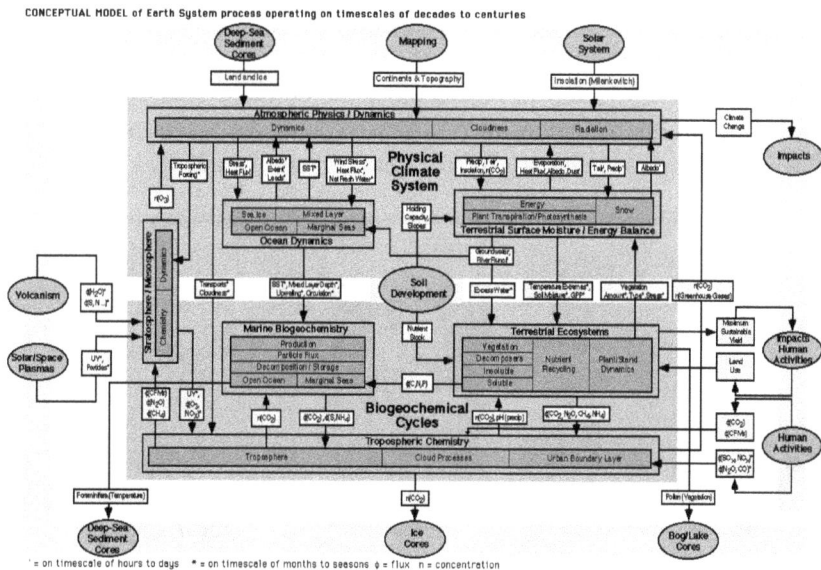

Fig. 1.5. Bretherton model for earth system processes.

A sensor unit that consists of a multi-measurement in-ground probe which incorporates a cluster of optical fibers, supplemented by a unique sensor design that integrates electrical wires within optical fibers ("wibers") represents the integrating technologies that may be used for advanced sensing. This multi-functional vertical probe clusters provides a platform for multiple, complimentary measurement techniques including: 1) Optical UV-VIS-IR-NIR absorption spectroscopy; 2) Optical thermometry; 3) Optical fluorescence spectroscopy; 4) Raman spectroscopy, and 5) Electrical impedance spectroscopy, including the capability of extraction of small aqueous samples. This suite of measurements collectively enables the simultaneous observation of gaseous species concentrations, temperature distributions, dissolved organic matter, and availability of nitrogen containing compounds at multiple depths within a soil column. The probe is augmented with a multiprotocol wireless communication module to allow for deployment in remote locations with appropriate levels of secure, net-centric data access.

In order to perform laboratory-based performance validation of the multisensor probe, a peat bog column was designed and constructed. The test "chamber" consisted of a 2 m optically transparent cylinder which incorporated an integrated sensing and ambient condition control system. The ambient condition control system allowed for varying a temperature and water content gradients throughout the column. Peat bog samples taken from northern Minnesota (USA) were loaded into the test chamber. The instrumented and controllable column is shown in Fig. 1.6.

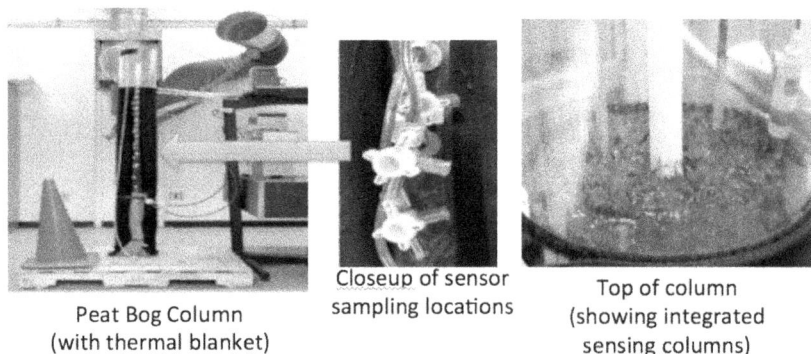

Peat Bog Column
(with thermal blanket)

Closeup of sensor
sampling locations

Top of column
(showing integrated
sensing columns)

Fig. 1.6. Peat bog column.

The peat samples were collected from the bog, transported to the laboratory, and placed into the peat bog column (see Fig. 1.7). A vertical temperature profilometry was immediately conducted using the sensor suite integrated into the column.

A fluorescent oxygen sensing probe was designed to provide the means with which the oxygen concentration at various depths within the peat bog itself could be measured. The technical challenge tackled involved the design and development of a chemical sensor which would map a variation in oxygen-level dependent wavelength/temporal intensity into an optical signal. This optical measurement of oxygen concentration – at specific localized depths within the peat bog column – would then be implemented using fiber optics and associated photonic devices.

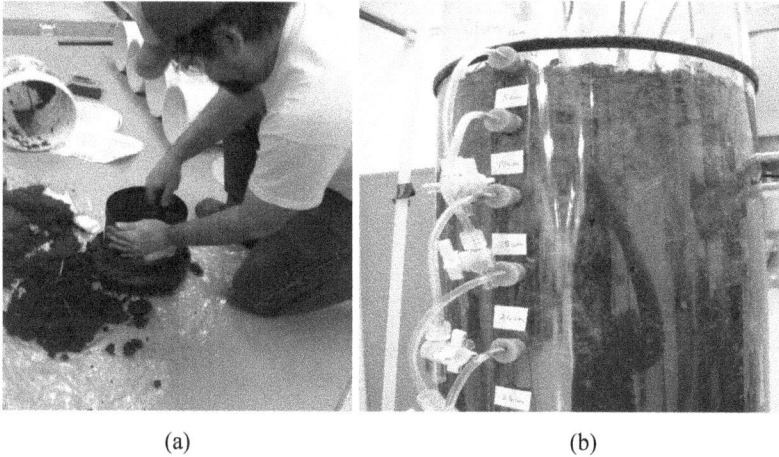

(a)   (b)

Fig. 1.7. Removing the peat samples from buckets prior to in column (a). The gas holes (for gas extraction analysis) are at 5 cm separations (b).

The fluorescent oxygen sensing probe relies on a chemical composition that changes fluorescence decay time as a function of oxygen concentration. In addition, the intensity of the fluorescence also varies with oxygen concentration level. A wide variety of configurations – and variations in chemical constituent levels – were evaluated. A photo montage of some of the sensor core components are shown in Fig. 1.8.

Fig. 1.8. The oxygen concentration sensors change intensity and temporal decay time (when subjected to a short activation optical pulse at an appropriate absorption wavelength).

As previously mentioned, the intensity as well as the exponential decay time of the fluorescence varies with oxygen concentration. The situation is illustrated in Fig. 1.9.

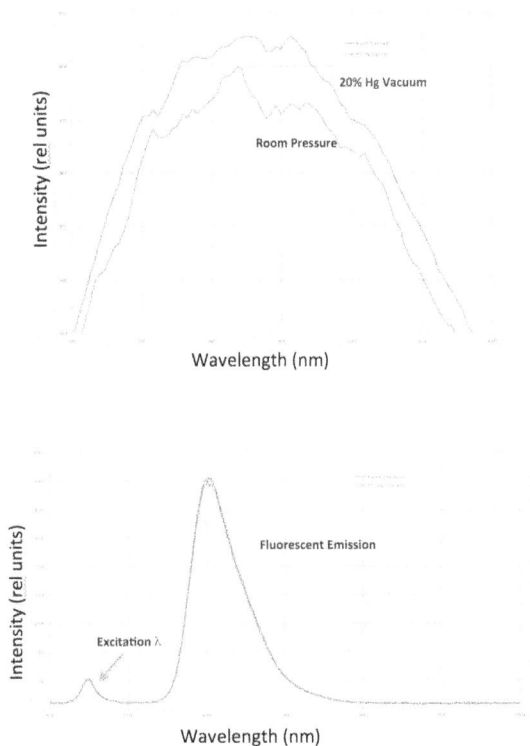

Fig. 1.9. The oxygen concentration sensors change intensity and decay (when subjected to a short activation optical pulse at an appropriate absorption wavelength).

In order to uniquely measure biological activity and associated chemistry at various depths, separation disks were designed that utilize and array sub-millimeter penetrations that allow sensors to pass through (see Fig. 1.10). The separation disks were treated with a super-hydrophobic coating that will not allow water transport across or around the disk, thereby preserving the unique chemistry of each region.

While field measurements were taken using the multifunction sensor suite, the primary reason for reviewing this activity within this Chapter is to highlight the complexity associated with the development and use of a single sensor probe. While deemed "successful" in terms of being able to measure what was proposed, this design is (essentially) not economically feasible for wide scale adoption and use.

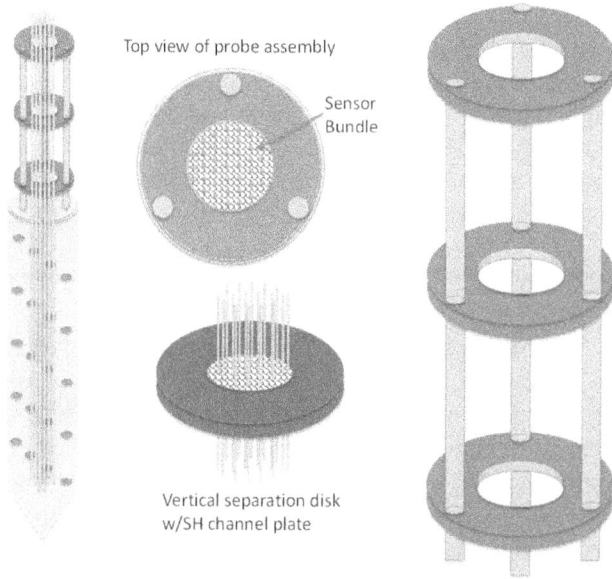

Fig. 1.10. Well point probe design with inherent structural integrity.

1.2.2. Sensing for Asset Monitoring and Fault Diagnosis in Electrical Systems

Review of the sensing needs within the electrical generation, transmission, and distribution system reveals that it is not just "voltage, current and transformer" status that are needed for optimal electric grid operation. Asset monitoring includes monitoring parameters that are indicative of the functional performance of the asset, monitoring the health condition of the asset, and diagnosing causes of abnormal performance or condition. Assets can include different types of generation, energy storage, and load as well as the electrical hardware of the power system (e.g., conductors, switches, reclosers, transformers, capacitors, voltage regulators, inverters) including emerging technologies. While many parameters of interest for these assets are physical (temperature, pressure, etc.), biological sensing (to a small extent) and chemical sensing (to a much larger extent) fall within the asset monitoring "needs".

Electrical grid assets will require monitoring to ensure that they are adequately performing their designed function and to properly maintain the asset to avoid unanticipated failure while in operation. A number of practical, operational benefits can be derived from increased measurement capabilities including: (1) Increased reliability and resilience through prevention of catastrophic failures of critical assets; (2) Delayed build-out of new transmission and other grid assets through more effective asset utilization; (3) More rapid detection and correction of critical fault conditions, and (4) Implementation of condition-based maintenance programs as a substitute for run-to-failure or time-based maintenance. These "needs" may be grouped into two logical categories: "Functional Performance" and "Health Condition" of the assets.

Functional Performance of Assets: As illustrated for a large electrical transformer, Fig. 1.11, the parameters to be monitored for function performance are largely electrical properties such as voltage, current, phase angle, and frequency. These parameters are used to calculate other power parameters such as real power, reactive power, harmonics, and power quality. Requirements for the accuracy, precision, and frequency of measurements is driven by how the information will be used for operation and maintenance of the power system. In some cases the usefulness of the measurement depends on how quickly the sensing and measurement data can be produced, integrated with other data, and converted to actionable knowledge. In other cases the technical precision and accuracy of the measurement will be dominant. In all cases, particularly for distribution assets, the cost of the sensor will play a critical role in dictating the potential for widespread deployment and hence the ultimate impact on the electrical grid infrastructure. In addition to electrical parameters, physical parameters such as position (open or closed) of switches, reclosers, breakers, fuses, etc., can also be considered related to functional performance of assets.

Health Condition of Assets: In some cases, the electrical parameters monitored for function performance can also provide useful information about the health condition of the asset. However, electrical parameters are often "lagging indicators" of the on-set of conditions for which asset maintenance is required. For this reason, monitoring alternative types of parameters can be more valuable for providing an early indicator of conditions for which timely maintenance can avoid impacts on function performance and extend the operational lifetime of assets. Examples of such measurements include chemical, mechanical, and thermal measurements and their changes with time, either abruptly or over extended time durations, which can reveal potential health issues of assets to enable condition-based maintenance programs. Prominent examples of parameters that fall within this category and their associated applications include the chemical changes that occur in the gases above and dissolved within the insulating oil for large power transformers as well as strain, temperature, and sag measurements on conductors as they stretch and contract with changes in weather and electrical loading conditions.

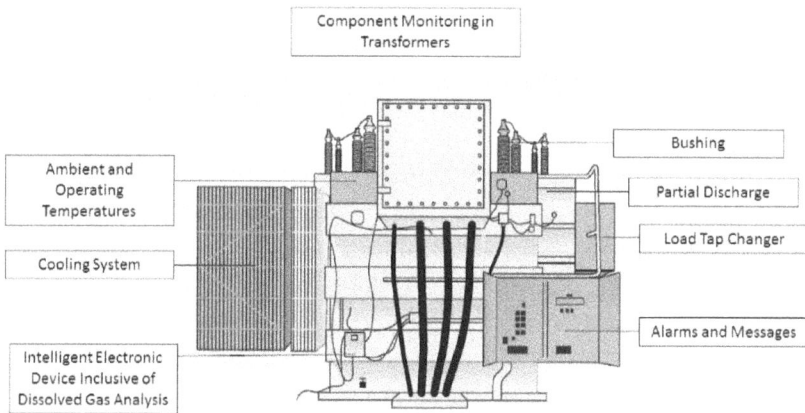

Fig. 1.11. Power transformer component monitoring schematic showing examples of instrumentation and measurements of interest.

A wide range of sensing and measurement technologies are currently employed for the purpose of monitoring grid assets, Table 1.1. One prominent example is dissolved gas analysis (DGA) techniques, which are commonly employed for diagnosing the operational health and condition of power transformers.

Such analysis is costly and time-consuming because it requires manual sampling and laboratory analysis techniques, but the benefits can still outweigh the costs for large power transformers, which are in operation for decades and represent major social, economic, and opportunity costs if they must be replaced due to unanticipated and often catastrophic failures.

Table 1.1. Key States and Parameters Relevant for Electrical Transmission and Distribution System Asset Monitoring and Fault Diagnosis.

States/Parameters	Directly measured or calculated from measurements	Sensors/meters required	Description and Note
Visual inspection	Direct observation	Photography, video monitoring, ultraviolet imaging	Deployed by drones, robots, manual
Temperature	Measured	Thermocouples and other point temperature sensors, IR imaging techniques	Point sensors identify single temperature locations while imaging tools such as IR imaging can also identify component hot spots while quantifying local temperature
Chemical analysis	Measured	DGA	Presence of N_2, O_2, H_2, CO_2, CO, CH_4, H_2, C_2H_6, C_2H_4, C_2H_2 in transformer insulating liquid or above the transformer oil in the gas phase
Tension	Calculated and Measured	Strain sensor, level or height monitoring	Can be inferred from direct strain sensors or indirect measurements such as line sag for transmission lines
Motion	Calculated	Vibration sensor, strain sensor, imaging based techniques	Camera imaging based methods with associated data analytics and motion proxies such as vibration or strain sensors.
Electrical equipment parameters	Calculated and Measured	Voltage and Current Transducers	Volt, current, phase angle, see previous sections on power flow and electrical grid state
Electrical discharge and corona	Calculated and Measured	Direct or calculated leakage current, local RF and static electric fields	Important for medium and high voltage energized assets including transformer bushings, transmission lines, etc.
Load tap changer position	Measured	Mechanical relay switch position	
Insulation oil level monitoring	Measured	Direct imaging or level indicator measurements	

In the case of particularly critical transformer assets, real-time diagnostic methods have been developed for on-line DGA, but they are far too expensive for widespread deployment. For lower voltage and power rated transformers such as distribution transformers, even conventional DGA analysis techniques become cost-prohibitive. As such, lower cost and robust sensing device solutions are of interest for real-time monitoring of the most important parameters associated with the dissolved gases including species such as H_2, CH_4, acetylene, ethane, ethylene, N_2, O_2 CO, CO_2, and others. In addition to DGA analysis, other sensors commonly employed for transmission transformers include bushings sensors, oil temperature and level, and tap position. Transformer bushings also represent a major source of catastrophic failure in transmission substations, and measurements of electrical parameters may be used to assess bushing health. Circuit breaker monitoring is another area in which existing technology solutions can be identified and can include gas temperature, pressure, and leak rate as well as mechanical systems.

A number of sensing and measurement technology platforms are currently under development to address needs in the area of asset monitoring. For example, the Electric Power Research Institute (EPRI) has developed a robust set of programs seeking to address sensing and measurement needs for transmission and substation applications. Examples of sensing technologies currently under development include: (1) RF sensors to monitor a broad range of relevant parameters for conductors including disconnections, fault currents associated with lightning, geomagnetic induced currents, temperature and inclination, motion and vibration, and proximity and tampering for security; (2) Optical diagnostic methods to monitor vibration and information about gas phase composition such as acetylene species surrounding high-temperature bushings; (3) Local hydrogen and other chemical composition sensors for power transformers; and (4) Unmanned aerial vehicle and robotic systems for inspection of overhead and underground transmission lines.

A key challenge associated with new and emerging sensing and measurement technologies required for asset monitoring applications is the need for compatibility with electrically energized components. In the case of applications within the distribution system, cost is also a key factor that will drive new technology development for new lower cost sensing solutions.

For geographically dispersed grid assets ranging from components to transmission and distribution lines, deployment of unmanned aerial vehicles instrumented with on-board sensing, imaging, or diagnostic capabilities or even with interrogation and data storage and management capabilities for interrogation of localized sensors show significant potential for wide area infrastructure monitoring. Similarly, application of satellite and wide area monitoring electromagnetic techniques, such as lidar and others, is anticipated to see increasing deployment moving into the future.

Having printed sensors on mobile platforms is a key "need".

1.2.3. Remote Sensing of CH_4 Using a Laser Tomographic Technique

A wide variety of remote sensing applications relying on optical retroreflectors – frequently with chemically sensitive coatings – have been designed, patented and deployed. In the case of identification of a source of CH_4 emission – such as finding the location of a pipeline leak – an array of retroreflectors may be dispersed around an area of interest. An accompanying laser, tuned to an absorption/emission wavelength of a chemical "of interest, is placed at a fixed location. The laser beam is then scanned from retroreflector to retroreflector with the accompanying detection system recording the received signal strength from each reflector. With each reading corresponding to the line-integral to/from laser-retroreflector, mathematical decomposition of the signals in a tomographic sense leads to a determination of the leak location. In addition, the retro-reflectors provide a means for the laser-based spectroscopy distance point measurements.

In recent years, laser-based standoff techniques have been developed for the detection of traces amounts of different materials [5, 6]. Despite the best efforts made, identification of these materials in a real environment remains a challenge. Many of these techniques rely on high power lasers that compromise the safety of the operator and bystanders, and their sensitivity and/or selectivity can be affected by environmental interferences. In order to overcome these challenges mid-IR laser based spectroscopy techniques are proposed.

Presently there exist many standoff detection methods. These techniques include Raman spectroscopy, laser induced fluorescence spectroscopy (LIFS), and laser induced breakdown spectroscopy (LIBS) [7-9]. Some

28

limitations of the laser based spectroscopic techniques are the fact that scattered photons need to reach the detector. To do so, they require high-power lasers, which are not eye safe and present a safety hazard for bystanders and/or the ecosystem including plants and animals. Besides the safety concerns, the complexity, high cost and bulky set-ups make them not suitable for field deployment or compact enough to make a human portable device. In addition, the Raman technique is a very weak effect, therefore it requires large amounts of sample for chemical identification. Techniques like LIBS and LIFS can cause damage to the sample under study. Therefore, standoff monitoring and characterization of the molecular signatures using these techniques is an extremely challenging task. These current techniques do not offer a clear path to develop a simple technology for remote, real-time GHG measurements suitable for standoff field measurements.

Why infrared (IR)? Molecules show specific activity in the IR spectrum (see Fig. 1.12). IR spectra of molecular bonds offer one of the most selective techniques for identifying molecular species. These spectra are characteristic to molecular groups and have an inherent selectivity unlike receptor-based molecular binding interactions used for speciation. Recent advances in eye-safe, tunable quantum cascade lasers (QCLs) in the mid-IR wavelength region enable larger standoff distances. The author – working with colleagues - has developed a laboratory-grade prototype for the longer wavelength region, 3.77-12.50 microns, and routinely obtain standoff spectra of chemicals adsorbed on surfaces [19-20]. Such prototype has been tested in the field at distances up to 25 meters, demonstrating the capability to interrogate various chemicals positively identifying them. This is due to the broad spectral range compare to single wavelength or narrow spectral range of other optical systems.

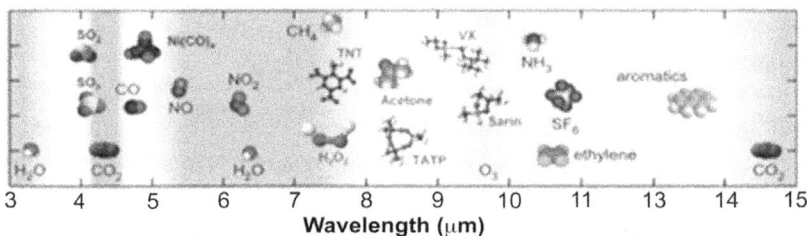

Fig. 1.12. Spectral activity for various molecules.

While the concept is easy, the equipment required to perform this task is expensive, beginning with a wavelength tunable quantum cascade laser (QCL). There are different types of retro-reflectors that can be utilized for laser-based distance measurement and object tracking over long distances. For example, corner-cube mirrors or prisms provide an almost perfect collimated return beam but have a limited angle of operation; the reflected power from a corner cube falls to -3 dB (50 %) of the value at normal incidence when the illumination is at an angle of 10° from the normal [25]. In the work reported by Grasso et al, [26] corner cube retro-reflectors are constructed from three mutually orthogonal reflective surfaces forming a concave corner. This mutual orthogonality ensures that light entering the corner reflector will be reflected back to the source, providing that it intercepts the corner reflector within its acceptance region. A ±60° acceptance angle was reported for the fabricated retroreflectors.

The same QCL lab prototype has been used to provide spatial information of environmentally important GHG emissions of interest, more specifically, CO_2 and CH_4. For these measurements, the information of specific absorption of the chemicals from many different optical absorption paths enabled a chemical tomographic reconstruction of the area using computed tomography algorithms. The combination these concepts allowed the first development of computed tomography standoff detection instrumentation for GHG over large areas with tomographic information [21].

To obtain the tomographic information of the field the infrared computed tomography prototype required a circular arrangement of mirrors to reflect the infrared light back to the source, as shown in Fig. 1.13. This mirror arrangement is not always practical in the field, considering that alignment of many mirrors can be a challenge and will require cleaning to keep maximum reflectivity due to exposure to the elements [24]. However, this technique demonstrated good sensitivity and reconstruction of a methane plume in a large field.

Another example of an optical standoff technique for field measurements is the Harris system GreenLITE, an optical system currently deployed in Paris for the 2016 Climate Action Summit [41-42]. Fig. 1.14, shows a picture of the system deployed in Paris, France. The GreenLite system provides information of the CO_2 concentration over a large area. The provides topological information of the distribution and concentration of the gas over the city by shooting a laser beam to retroreflectors at specific locations. The laser beams are reflected back to a detector where the attenuation of the light is measured and displayed real time.

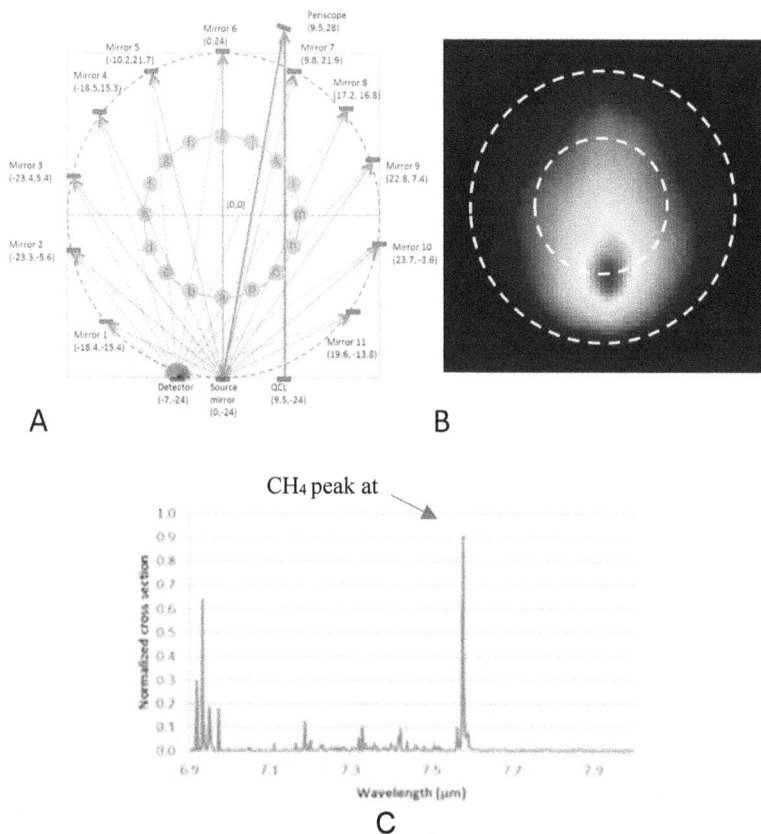

Fig. 1.13. Multispectral infrared computed tomography using quantum cascade lasers. (A) Set up schematics, (B) Infrared multispectral computed reconstruction for CH_4 plume, and (C) CH_4 infrared spectra extracted from the computed reconstruction.

Optical techniques are demonstrated to be highly selective and can be used for chemical detection over long distances with ppm/meter sensitivity. These techniques are very reliable, but when it comes to deployment in the field there can be challenges. Depending on the application, the practical utilization of this technique is problematic based in large part on equipment cost and system complexity.

Fig. 1.14. Harris GreenLite system deployed in Paris, France 2015. [42]

1.2.4. Large-scale Sensing

Numerous large scale "observation projects", such as the Arctic Observing Network (AON), the Ocean Observing Initiative (OOI), and the National Ecological Observatory Network (NEON), rely on sensors and systems to provide the measurement information upon which scientific studies and subsequent reports rely. The problems being addressed are complex yet require multiple parameters to be measured simultaneously as well as being collocated. Consider NEON, whose overall intent is presented in this extraction from a National Academy of Sciences report:

"The National Ecological Observatory Network (NEON) proposed by the National Science Foundation (NSF) would be a network of infrastructure that would support continental-scale research on pressing environmental challenges. Major Research Equipment and Facilities Construction (MREFC) funding was requested to build a network for a coordinated, nationwide multisite network for experimental and observational environmental research. NEON would enable the study of common themes and the sharing of data and information across sites. It would facilitate a more integrated

approach than merely linking existing research sites, such as the Long Term Ecological Research (LTER) sites, by allowing research on drivers of environmental changes to be pursued across the complete spectrum of ecosystems. It would be dedicated to producing the key results and fundamental scientific principles that are needed to project how human actions would likely affect natural and managed ecosystems across the nation in the coming decades."

The measurement needs associated with such lofty goals – of NEON and similar large scale observation networks, Fig. 1.15 - must be coupled to a technological "solution" that is cost effective, easy to implement or deploy and provide the basis for a wide range of parameters to be sensed.

Fig. 1.15. The NEON observation network has measurement requirements spanning multiple timescales and spatial resolutions.

1.2.5. Common Theme – Common Need

The application areas presented in this Section have one common theme: the need for low cost, easy to deploy sensors to measure chemicals, biological agents, (traditional) physical parameters. The current sensing technologies used in the vast majority of research associated with these areas are of a more traditional sense using systems that take the parameter(s) of interest and transduce it into another area, typically electrical.

1.3. Printed Sensors and Electronics

Highly sensitive and low cost sensors are needed for remote and continuous monitoring of green house gases emissions, methane leaks, or air pollution in the field. As evident in Section 1.3, growing concerns over climate change, leaks from hydraulic fracturing sites, adverse effects of air pollution on public health and potential explosions from aging natural-gas pipes are drives for rapid, and/or portable detection system for detection, identification and continuous monitoring of these gases at trace levels [21, 24, 29-31]. Current methods for detection of greenhouse gas (GHG), more specifically methane, include gas chromatography with flame ionization detection (GC/FID), long wavelength optical thermography, and sampled gas Fourier Transform IR spectroscopy (FTIR). Contrary to optical methods, GC/FID doesn't offer the capability of remote and/or continuous monitoring of GHG in the field, while the thermography of CH_4 relies on other molecules not radiating in the same wavelength region.

The methods described above offer the sensitivity and selectivity desired to perform monitoring of different parameters in the field. However, low cost, light weight and continuous monitoring of single of multiple parameters in the field is still a challenge. Advancements in additive manufacturing and materials have made it possible the evolution of printed electronics. Printed sensors are presented as an alternative to perform real time field measurement at a low cost. Fabrication methods using additive manufacturing techniques – 3D printing – coupled with a sensor-interrogator design that allows different parameters to be measured in a non-contact means, via a radio signal. This form of printed electronics is specific to a sensor design with no active electronics at the sensor itself thereby making it intrinsically safe. The practical ramifications for such a design and system is that the sensor may be deployed in explosive or hazardous environments, is interrogated from a distance (via the radio signal), and may be very inexpensive to manufacture. The technological basis for the sensor itself is a variant on classic optical interferometry – with a reference and a sensing "arm" – but fabricated on a substrate that allows a surface acoustic wave to propagate with nominal loss and pulse dispersion.

Advancement in manufacturing and printed electronics have opened the door to the development of new sensing technologies and the option of mass production. Applications in industries like medicine, food [39], textiles as in wearables electronics and energy will benefit of the financial aspects of printed sensors. Fig. 1.16 illustrates examples of printed sensors for different applications. Take, for example, smart packaging, were printed RFID tags are now found in medicine and food for traceability. In this case the RFID tags is a label containing a chip/sensor combination capable to record the time and temperature history of the product [40]. This technology allows the consumer to have real time information on the quality of the product prior to buying. Other technologies were printed electronics are being implemented are, health monitoring, communications, and environment.

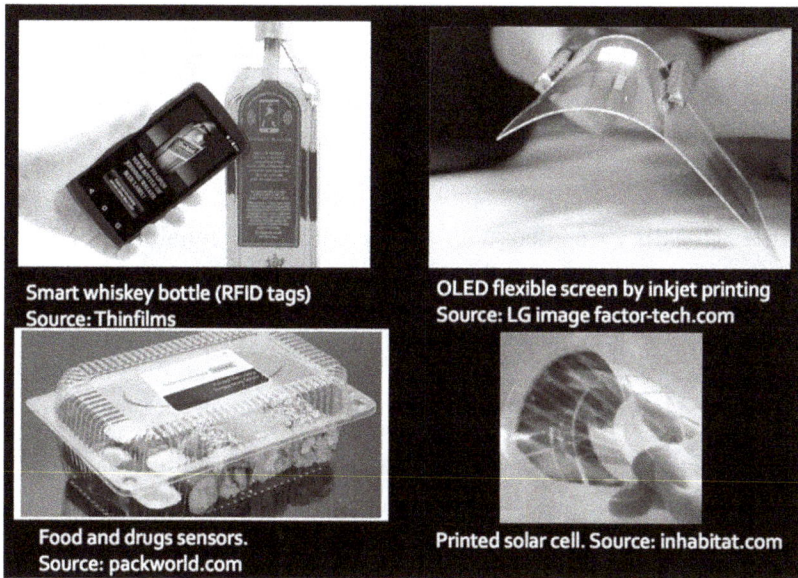

Fig. 1.16. Printed sensors and devices.

The work published by Hayat et al. [41] discusses the developments in screen printing technologies to develop small and disposable electrochemical sensors. Such sensors are used for environmental monitoring like heavy metal detection, water quality, organic compounds and bacterial for example. The detection was achieved by special coating of the electrodes.

Another example where printed electronics was used to develop environmental sensors is the work published by Crowley et al. [42]. In this work, a printed ammonia sensor was developed by inkjet and screen printing. Screen printing was used to print the silver interdigitated array (IDA) and inkjet printing to deposit the polyaniline nanoparticle suspension. Polyaniline is a conductive polymer that reacts with in contact with ammonia, making it suitable for development ammonia sensors. The results show that the sensitivity to ammonia was in the range of 1-100 ppm.

1.3.1. Passive Wireless Sensors

An ideal example for passive wireless sensors is the surface acoustic wave (SAW) technology. SAW offers a broad range of sensor applications, used by different industries, for example, medical, aerospace, telecommunications, industrial, commercial and automotive [1-3]. These are used as bandpass filters, pressure, humidity, and temperature or chemical sensors, just to mention a few applications. Some of these applications requirements are very challenging such as low cost, maintenance free, self-powered, long lasting, light, small, self-cleaning and able to operate under extreme conditions. These advantages of the versatility coupled with the demonstrated reliability of SAW devices leads to applying this technology to the remote sensing and continuous monitoring of greenhouse gases emissions in the field with enough sensitivity to include biogenic sources. In practice, the SAW devices are positioned at different locations in the field providing specific and localized information of the concentration for a specific gas.

Traditionally SAW devices are fabricated in several stages in a microfabrication, clean room facility using complicated procedures, like lithography, that include special preparation of the piezoelectric crystals, masks for electrode deposition [4-14]. Microfabrication facilities costs millions of dollars including maintenance, materials and operational costs. Such facilities produce millions of SAW devices a year reducing their overall cost. Although, the overall cost of SAW devices is low, customizing them to a specific application can increase the price of the end device.

An alternative to fabrication of these devices is by printed electronics. Currently there are different systems that can provide with the resolution and precision needed for fabrication and customization. Printed electronics manufacturers have improve the deposition method for these systems. For example, the aerosol jet systems provides the ability to print fine-feature electronic, structural, onto almost any surface, with feature sizes in the range of 1mm to as small as 10 microns. Such systems are ideal for the development of biochemical sensors since allows the use of any ink, including biological agents for deposition on a surface. The printing process is a non-contact and conformal, allowing for patterning over existing structures, across curved surfaces, and into channels.

To write conductive patterns on surfaces, research activities abound investigating the performance of metallic nanoparticle based inks used for printing. Such studies reveal the ideal conditions, such as curing time, deposition rate, nanoparticle size, among others, for best conductivity and device performance. The net result of those studies is to provide materials allowing for the direct printing of sensors for biochemical environmental and physical parameter measurements.

An illustrative diagram of a printed electronic, passive, SAW device with biochemical sensing is presented as Fig. 1.17. Please note that other transducer-reflector designs are also used.

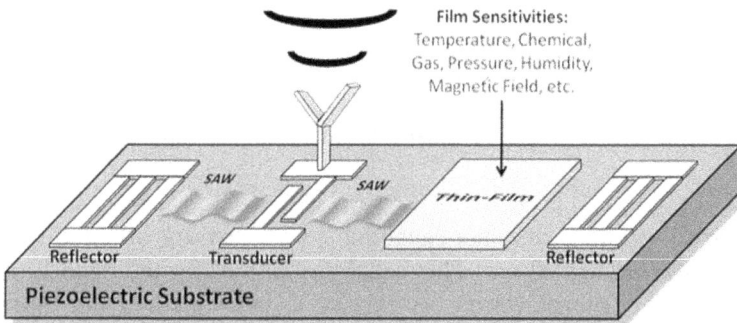

Fig. 1.17. Printed electronic sensor, a SAW device.

Substrate materials for SAW devices include ferroelectric polymer (PVDF), Lead Zirconate Titanate (PZT) and Lithium Niobate (LiNbO$_3$). Typical SAW devices operational frequency range is between 5-500 MHz. The resonance frequency is determined by

$$f_0 = \frac{V_R}{4d}, \qquad (1.1)$$

where V_R is the velocity of the crystal, d is the finger and gap width. This frequency dictates the interrogator's operating frequency. Fig. 1.16 shows a basic design showing the direct printed interdigitated transducer (IDT), electrodes, and antenna.

In an operational system, the direct printed sensor is a passive device powered by the interrogator signal. Sensor readings are obtained by the interrogator via a transmitted radio waves to the sensor and an associated will returned signal, Fig. 1.18.

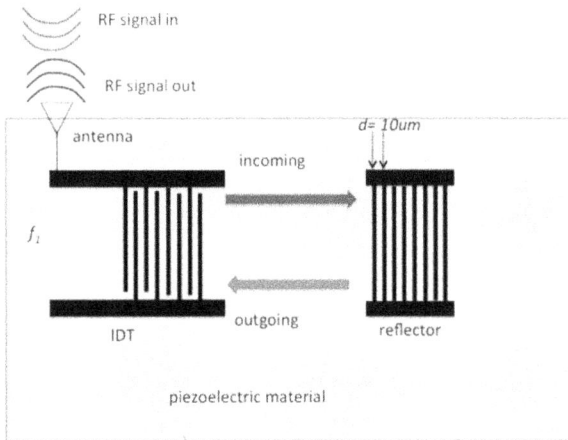

Fig. 1.18. Basic design of a direct printed interdigitated transducer (IDT).

An aerosol printer offers the capability to use any ink or solution with particle size up to 500 nm. This capability allows for the deposition of films on the piezoelectric crystal as a new method for surface functionalization. Consider the case where the target is to have a sensor platform able to measure CO_2 and CH_4 from biogenic sources in the environment. Recent developments have demonstrated that SAW sensors coated with graphene–Ni–L-alanine composite films to detect CO_2 are effective with a detection limit of 200 ppm [1]. Other CO_2 sensitive film, like polyimides or the fluoropolymer Teflon AF 2400TM have demonstrated to selectively adsorbs large quantities of CO_2 with sufficient diffusivity to allow response times of a few seconds to be achieved [3].

Other films have been studied to develop a new methane gas sensor incorporating SAW technology and cryptophane A (CrypA films), which try to provided fast and accurate sensor with detection limits of 0.2 % - 0.5 % and 204 Hz/%. The deposition of this film on the piezoelectric crystal was done from solution by drop-coating [14]. There are a number of studies found in the literature that demonstrate sensitivity and selectivity of tin oxide SnO_2 thin films to methane. These studies report the sensitivity of these films to be in the range of 200-1,000 ppm [15-27]. In each case, independent of "sensing element" chemistry, the overall printed electronic SAW design remains the same, with the interrogation signal and system functioning the same, namely that a change in received signal frequency and/or amplitude indicates a change in the parameter being measured.

In a typical instantiation, the sensitive films have a thickness in the range of (100-500 nm). Correspondingly, the film's surface functionalization is deposited by *rf*-sputtering from a single mixed-phase oxide target. Therefore, appropriate processing protocols are used to create a porous nanostructured surface morphology on piezoelectric substrates. The porous layer allows gas species to fully penetrate the structure, rather than be confinement only on the surface. Hence, as shown in Table 1.2, the sensitivity of the sensor significantly increases because the gas sensing reactions take place over the entire volume of the sensing layer.

Table 1.2. Selective coatings for printed devices.

Material	Deposition Method	Gas
Graphene–Ni–L-alanine[1]	Electrochemical	CO_2
Carbon Nanotubes-PEI	Drop Coating	CO_2
Teflon 2400	Coating	CO_2
Indium tin oxide	Sputtering*	NO_2
Cryptophane A	Drop Coating	CH_4

In addition to traditional surface modification methods like, rf-sputtering, other direct printing methods may be used. As previously mentioned, Wang et. al. [14], demonstrated film deposition using drop-coating methods. When using a suitable aerosol jet printer, a solution formulation may be developed as an ink and then used in direct printing for direct coating on the SAW surface. This process eliminates extensive and complex coating procedures reducing the cost of the fabrication of the final device.

When a SAW sensor is exposed to the target gas, its electrical and mechanical characteristics change according to the absorption capacity of the active sensing region, resulting in a frequency shift. The change in frequency caused by mass loaded on the SAW surface is defined by,

$$\Delta f = (k_1 + k_2) f_0^2 \sigma, \qquad (1.2)$$

where k_1 and k_2 are the coupling constants determined by different displacement components of the SAW in the substrate, f_0 is the unperturbed oscillating frequency of the SAW, and $\sigma = hp$, h is the thickness of the gas layer and p its density. This equation assumes that the absorbed gas forms an isotropic, non-piezoelectric, non-conducting layer [28-34]. It is this frequency shift utilized in the quasi-interferometric SAW sensor design that allows the interrogator to determine the chemical concentration level.

1.3.2. Advanced Printed Sensors: SAW and Retroreflectors

A further refinement to the retroreflector sensing method as well as the SAW sensor method is to combine aspects of each into a specific SAW device with retroreflectors integrated onto the SAW device surface.

Miniature, < 5 mm, retro-reflectors for mid and long wave infrared portion of the spectrum have been developed using a variety of production methods. These include low-cost production methods such as for spherical retro-reflectors gradient index glass, glass compression molding, injection-molding technologies, and fluidized bed chemical vapor deposition (CVD). Spherical retro-reflectors are ideally isotropic and can be capable of extremely large acceptance angles, limited only by the effects of spherical aberration and available construction methods. The fabrication of a mid and long wave infrared miniature spherical retro-reflector has been demonstrated [27]. The performance of the retro-reflector is quantified by the scattering cross section (CS). The CS is one performance metric that describes both how much of the incident radiation is intercepted by the retro-reflector and how efficiently the radiation is redirected toward the source [28].

$$CS = \lim_{\to D} \left(\frac{\pi^3}{4\lambda^2}\right) d^4 S(d) \, , m^2, \qquad (1.3)$$

where D is the diameter of the sphere, and S(d) is the Strehl ratio computed for an exit pupil of diameter d. The Strehl ratio in this context is the ratio between the actual intensity on the optical axis and the intensity that would be produced by a mirror of the same aperture with a reflectivity of 100 %.

An integrated retroreflector-SAW bio/chemical sensor design is presented as Fig. 1. 19.

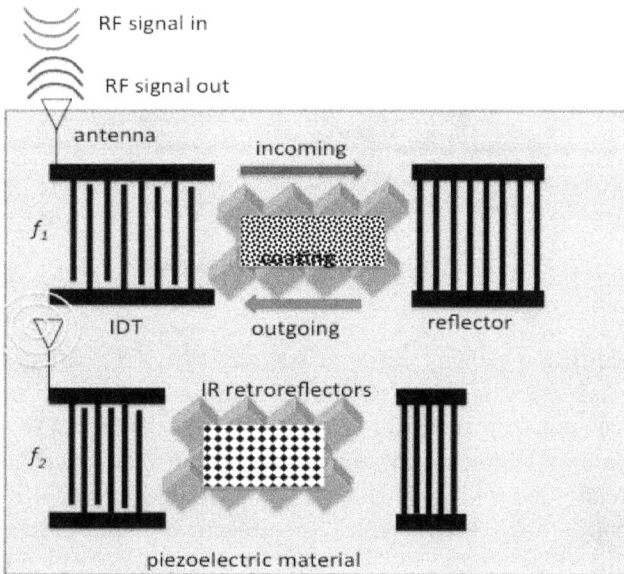

Fig. 1.19. SAW device platform with chemical coating on IR retroreflectors.

In the case of enhanced CH_4 sensing, commercially available retro-reflectors of different sizes, 7-50 mm, are chosen based on their reflectivity at the wavelength region of interest - 3.77 to 12.50 microns. The combined optical-SAW printed sensor requires that the retro-reflectors be attached between the IDT's and/or RF reflectors on the SAW device, as illustrated in Fig. 1.20. Note that it is at this location where the chemical coating selective to the target species, for example, CO_2 and CH_4, is deposited.

The result of such a design is a method of increasing the sensitivity only when the QCL beam illuminates the chemical coating. Line integral and the point sensing measurement provide a wealth of information related to the exact reading at the location of the deployed sensor(s) as well as the area between the sensors (as retroreflectors) and the QCL and detector.

Fig. 1.20. Fabricated sensors.

Laser spectroscopy using infrared wavelength QCLs offers compact, tunable, and eye safe, mid-IR sources with sufficient power for non-destructive standoff detection in combination with SAW chemical sensing. We will combine these two technological advances to develop a novel technique taking advantage of the molecular specificity of IR spectroscopy for long path spectroscopic measurements and functionalized SAW devices for localized chemical sensing approach to monitor GHG in the field. This new technique will be capable of obtaining molecular signatures of GHGs with high sensitivity and selectivity of a localized region by the SAW devices and along multiple paths, up to kilometer distance, by the multispectral laser spectroscopy.

Consider a system where the QCL-based standoff detection system will provide single measurements of GHGs along a path that crosses the area of interest using a laser and detector on one side of the area and a functionalized SAW device on the opposite side. The laser light irradiates the multiple SAW sensors in the field, individually, to collect data. The embedded retroreflectors on the SAW devices reflect the light back to the laser system where it is collected by a telescope and focused onto the detector. The irradiation of the laser wavelengths on the SAW provides specific absorption behavior for the gas absorbed on the surface, increasing the selectivity. At a wavelength associated with absorption of a particular gas, the response from such a system follows Beer's law:

$$I = I_0 e^{-\Sigma_i \mu_i \rho_i l_i}, \qquad (1.4)$$

where I is the measured intensity, I_0 is the transmitted intensity with no gas, μ_i is the cross section of the gas at the measured wavelength, ρ_i is the gas density, and l_i is the path length through the gas. To obtain the absorption of the gas most optical spectroscopy sensors incorporate a reference gas cell with known gas concentration. For this application the SAW device placed at a specific location provides concentration information of the particular gas at that location and serves as a reference concentration to calibrate the QCL open path absorption by the gas.

A schematic diagram of the integrated system is represented in Fig. 1.21.

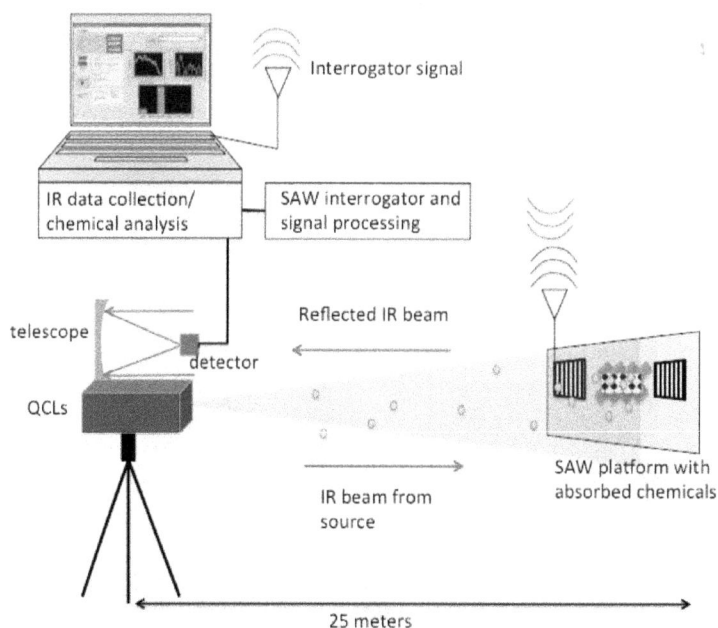

Fig. 1.21. Schematic diagram of free-space IR spectroscopy with RF interrogation of a SAW+retroreflector sensor.

The SAW sensors are positioned at different locations to obtain information about the location of the leak or gas emissions and gas concentration. The RF interrogator, signal emitter of the SAW is embedded with the laser system. The sensitivity of SAW devices is designed to be in the range of hundreds of ppms depending of the performance of the chosen coating and coating procedure. Remote excitation of the SAW devices by optical probing is an interesting area to explore due to recent advances on QCL sources [23]. Optical probing of the SAW devices make the entire system more compact as it only needs lasers sources for SAW excitation and spectroscopic analysis.

1.4. Conclusions

Printed electronics, in general, and printed electronic sensors in particular is an ever expanding field. The capability of using additive manufacturing tools and techniques in conjunction with novel sensor designs, such as the passive wireless biochemical sensors described in this Chapter, leads to a wide range of possibilities in terms of low cost, easy to deploy devices. One such application is the detection of fugitive emissions from storage tanks, such as those shown in Fig. 1.22.

Fig. 1.22. Mobile platforms with embedded printed electronic biochemical sensors may be used for standoff inspection, such as this hydrofracking storage battery near Midland Texas.

Statements at recent technology symposia reveal that printed electronics directly intersects the Internet of Things (IoT). A specific reason is that silicon-based sensors are the first that have been associated with IoT technology yielding a wide array of sensors with multiple "internet connected device" applications, such as track data from airplane, biochemical sensing of food, wind turbines, engines, and medical devices.

However, these silicon-based are not suitable for several other applications. Bendable packaging and premium items are some of the application where embedded sensors may not prove optimal while printed electronics may be. Using sensor technology, information is transferred on smart labels that can be attached to packages to be tracked in real time. Being able to print electronic sensors onto various material substrates, having a design that is intrinsically safe, requires no connected power source, is inexpensive and easy to interrogate will also allow for use in mobile platform use, specifically unmanned aerial systems (aka, drones).

References

[1]. Sheng Xu, Cuiping Li, Hongji Li, Mingji Li, Changquing Qu, and Baohe Yang, Carbon dioxide sensors based on a surface acoustic wave device with a grapheme-nickel-L-alanine multilayer film, *J. Mater. Chem. C*, 3, 2015, p. 3882.
[2]. Donald C. Malocha, Mark Gallagher, Brian fisher, James Humphries, Daniel Gallagher and Nikolai Kozlovski, A passive wireless multi-sensor SAW technology Device and System Perspectives, *Sensors*, 13, 2013, p. 5897.
[3]. Bill Drafts, Acoustic/Ultrasound Acoustic Wave Technology Sensors, *Sensors Online*, 2000.
[4]. G. Potter, N. Tokranova, A. Rastegar, and J. Castracane, Design, fabrication, and testing of a surface acoustic wave devices for semiconductor cleaning applications, *Microelectronic Engineering*, 162, 2016, p. 100.
[5]. M. R. Zakaria, U. Hashim, R. Mat Ayub, and Tijjani Adam, Desing and Fabrication of IDT SAW by Using Conventional Lithography Technique, *Middle-East Journal of Scientific Research*, 18, 9, 2013, p. 1281.
[6]. Soo-Hyung Seo, Wan-Chul Shin, Jin-Seok Park, A novel method of fabricating ZnO/diamond/Si multilayers for surface acoustic wave device applications, *This Solid Films*, 416, 2002, p. 190.
[7]. Morales-Rodriguez, M. E., Senesac, L. R., Rajic, S., Lavrik, N.V., Smith, D.B. and Datskos, P. G., Infrared Microcalorimetric Spectroscopy using Quantum Cascade Lasers, *Opt. Lett*, 2013, 15, pp. 507-509.

[8]. Van Neste, C. W., Morales-Rodriguez, M. E., Senesac, L. R., Mahajan, S. M. and Thundat, T., Quartz Cyrstal Tuning Fork Photoacoustic Point Sensing, *Sensors and Actuators B: Chemical*, 105, 1, 2010, pp. 402-405.

[9]. C. W. Van Neste, L. R. Senesac and T. Thundat, Standoff Spectroscopy of Surface Adsorbed Chemicals, *Analytical Chemistry*, 81, 1952, 2009.

[10]. C. W. Van Neste, L. R. Senesac and T. Thundat, Standoff photoacoustic spectroscopy, *Applied Physics Letters*, 92, 2008, 234102.

[11]. Es-Sebbar and Et-Touhami, Absolute nitrogen atom density measurements by two-photon laser-induced fluorescence spectroscopy in atmospheric pressure dielectric barrier discharges of pure nitrogen, *Journal of Applied Physics*, 106, 2009, 073302.

[12]. L. J. Radziemski and D. A. Cremers, Handbook of laser-induced breakdown spectroscopy, *John Wiley*, New York, 2006.

[13]. J. Moros, J. A. Lorenzo, P. Lucena, L. M. Tobaria, and J. J. Laserna, Analysis of explosives using a mobile integrated sensor platform, *Anal. Chem.*, 82, 2010, pp. 1389-1400.

[14]. Wen Wang, Haoliang Hu, Xinlu Liu, Shitang He, Yong Pan, Caihong Zhang, and Chuan Dong, Development of a room temperature SAW methane gas sensor incorporating a supramolecular cryptophane A coating, *Sensors (Basel)*, 16, 2016, 73.

[15]. Zenghui Wang, Peng Wang, Jaesung Lee, Chung-Chiun Liu, and Philip X. L. Feng, Towards real-time methane (CH$_4$) capture and detection by nanoparticle-enhanced silicon carbide trampoline oscillators, in *Proceedings of the 18th IEEE International Conference on Solid-State Sensors, Actuators and Microsystems (TRANSDUCERS'15)*, 2015, 432.

[16]. Divya Haridas, Vinay Gupta, Enhanced response characteristics of SnO$_2$ thin film based sensors loaded with Pd clusters for methane detection, *Sensors and Actuators B*, 156, 2012, pp. 166-167.

[17]. V. K. Yatsimirskii, N.P. Maksimovich, A. G. Telegeeva, N. V. Nikitina, and N. A. Boldyreva, Semiconductor sensors based on SnO$_2$ with Pt additives and their catalytic activity in oxidation of methane, *Theoretical and Experimental Chemistry*, Vol. 41, Issue 3, May 2005, pp. 187–191.

[18]. Jae Chang Kim, Hee Kwon Jun, Jeung-Soo Huh, Duk Dong Lee, Tin oxide-based methane gas sensor promoted by alumina-supported Pd catalyst, *Sensors and Actuators B*, 45, 1997, p. 271.

[19]. C. W. Van Neste, L. R. Senesac and T. Thundat, Standoff Spectroscopy of Surface Adsorbed Chemicals, *Analytical Chemistry*, 81, 2009, p. 1952.

[20]. C. W. Van Neste, L. R. Senesac and T. Thundat, Standoff photoacoustic spectroscopy, *Applied Physics Letters*, 92, 2008, 234102.

[21]. Phillip Bingham, Marissa E. Morales-Rodriguez, Panos Datskos, and David Graham, Multi-spectral infrared computed tomography, in *Proceedings of the IS&T International Symposium on Electronic Imaging Computational Imaging XIV*, 2016.

[22]. L. Reindl, G. Scholl, T. Ostertag, C. C. W. Ruppel, W. E. Bulst, and F. Seifert, SAW devices as wireless passive sensors, in *Proceedings of the IEEE Ultrasonics Symp.*, 1996.

[23]. George I. Stegeman, Optical probing of surface waves and surface wave devices, *IEE Transactions on Sonics and Ultrasonics*, 23, 1, 1976.

[24]. Kang Sun, Lei Tao, David J. Miller, Mark A. Zondlo, Kira B. Shonkwiler, Christina Nash, and Jay M. Ham, Open-path eddy covariance measurements of ammonia fluxes from a beef cattle feedlot, *Agricultural and Forest Meteorology*, 213, 2015, p. 193.

[25]. V. A. Handerek and L. C. Laycock, Feasibility of retroreflective free-space optical communication using retroreflectors with very wide field of view, in Advanced Free-Space Optical Communications Techniques and Technologies, M. Ross and A. M. Scott (Eds.), *Proc. SPIE*, 5614, 2004, 1.

[26]. Robert J. Grasso, Jefferson E. Odhner, Hamilton Stewart, Robert V. McDaniel, Laser radar range and detector performance for MEMS corner cube retroreflector arrays, *Proc. SPIE Advanced Free-Space Optical Communications Techniques and Technologies*, 5614, 2004, p. 43.

[27]. Bruce E. Bernaki, Norman C. Anheier, Kannan Krishnaswami, Bret D. Cannon, and K. Brent Binkley, Design and fabrication of efficient miniature retroreflectors for the mid- and long-wave infrared, *Proc. SPIE*, 6940, 2008, 1.

[28]. John P. Oakley, Whole-angle spherical retroreflector using concentric layers of homogeneous optical media, *Applied Optics,* 46, 2007, p. 1026.

[29]. Patricia Daukantas, Air-Quality monitoring in the mid-infrared, *Optics and Photonics News*, 2015.

[30]. D. J. Miller, K. Sun, L. Tao, M. A. Khan, and M. A. Zondlo, Open-path, quantum cascade-laser-based sensor for high-resolution atmospheric ammonia measurements, *Atmos. Meas. Tech,*. 2014, 7, 81.

[31]. Guidelines for the measurements of methane and nitrous oxide and their quality assurance; Global atmosphere watch report No. 185 (WMO/TD-No. 1478); *World Meteorological Organization*, Geneva, Switzerland, 2009.

[32]. Rabih Maamary, Xiaojuan Cui, Eric Fertein, Patrick Augustin, Marc Fourmentin, Dorothee Dewaele, Fabrice Cazier, Laurance Guinet, and Weidong Chen, A quantum cascade laser-based optical sensor for continuous monitoring of environmental methane in Dunkirk (France), *Sensors*, 16, 2016, 224.

[33]. Wohltjen, H. Mechanism of Operation and Design Considerations for Surface Acoustic Wave Device Vapor Sensors, *Sens. Actuators,* 5, 1984, pp. 307–325.

[34]. Jiansheng Liu, and Yanyan Lu, Response mechanism for surface acoustic wave gas sensors based on surface-adsorption, *Sensors*, 14, 2014, 6844.

[35]. L. Reindl, G. Scholl, T. Ostertag, C. C. W. Ruppel, W. E. Bulst, and F. Seifert, SAW devices as wireless passive sensors, in *Proceedings of the IEEE Ultrasonics Symp.,* 1996.

[36]. George I. Stegeman, Optical probing of surface waves and surface wave devices, *IEE Transactions on Sonics and Ultrasonics*, 23, 1, 1976.

[37]. Yizhong Wang, Minking K. Chyu, Qing-Ming Wang, Passive wireless surface acoustic wave CO_2 sensor with carbon nanotube nanocomposite as an interface layer, *Sensors and Actuators A*, 220, 2014, pp. 34-44.

[38]. Chunbae Lim, Wen Wnag, Sangsik Yang, Keekeun Lee, Development of SAW based multigas sensor for simultaneous detection of CO_2 and NO_2, *Sensors*, 15, 2015, pp. 30187–30198.

[39]. Schroeter, K., Printed sensors: enabling new applicatons, *Sensors Review,* 28, 1, 2008, pp. 6-11.

[40]. Harrop, D. P., Printed electronics in supermarkets, *IDTechEx*, 30 March 2009.

[41]. Akhtar Hayat, and Jean Louis Marty, Disposable screen printed electrochemical sensors: tools for environmental monitoring, *Sensors,* 14, 6, 2014, pp. 10432-10453.

[42]. Karl Crowley, Aoife Morrina, Aaron Hernandeza, Eimer O'Malleya, Philip G. Whittenb, Gordon G. Wallaceb, Malcolm R. Smytha, Anthony J. Killard, Fabrication of an ammonia gas sensor using inkjet-printed polyaniline nanoparticles. *Talanta*, 77, 2, 2008, pp. 710–717.

[43]. United Nations Development Goals, *UN Climate Change Conference Paris*, http://www.un.org/sustainabledevelopment/cop21/ (accessed March 2017)

[44]. Green Lite Ground Remote Sensing, *Harris-Technology to Connect, Inform and Protect*, https://www.harris.com/solution/greenlite-ground-remote-sensing (accessed March 2017).

Chapter 2

Development and Applications of Upconversion Nanoparticle in Biosensing

Liuen Liang, Andrei V. Zvyagin and Run Zhang

2.1. Introduction

Biochemical sensing is of growing importance in healthcare applications due to the capability of early detection of disease biomarkers as well as real-time and continuous monitoring of target analytes in complex biological activities. Among the various sensing techniques, optical sensing using fluorescent or luminescent biosensors for the detection of trace amounts of analytes is getting increasing attention because of its simplicity, high sensitivity, and versatility in obtaining diverse information in biological systems. The applications of these optical biosensors in non-invasive detection of analytes in live cells, tissues, and animals are also speeding up with the supports from rapid development of optical microscopy.

In light of that, the use of fluorescence receptors such as organic dyes, transition metal complexes, rare-earth chelates, and quantum dots has attracted considerable interest in the sensing research community. However, *in vitro* and *in vivo* assays of biomarkers remains challenges due to high activities of analytes and lacks of effective detection approaches. In particular, development of fluorescent and/or luminescent biosensors for the detection of highly reactive biomolecules in

Run Zhang
School of Chemical Engineering, University of Science and Technology Liaoning, 185 Qianshan Zhong Road, Anshan 114044, P. R. China
Australian Institute for Bioengineering and Nanotechnology, The University of Queensland, St Lucia, QLD, 4072, Australia

49

complicated biological conditions is highly demanded. Typically, desirable optical biosensors must: (i) Fast respond to biomolecules, giving changes in fluorescence/luminescence signals; (ii) Be highly sensitive and selective towards targets; (iii) Discriminate the background autofluorescence from biological samples; (iv) Be highly photostable, ensuring their long-term *in vivo* monitoring of biological activities; (v) Possess low toxicity and exert no phototoxicity to live bodies during detection.

These criteria are satisfied when using near-infrared (NIR)-excitable nanoparticles, where the luminescent nanoparticles can be excited by NIR light to elicit the photoluminescence for biological analysis. NIR light falls into the biological transparency window (700-1300 nm), it penetrates deeper and excites considerably less autofluorescence and exerts less phototoxicity compared to UV and visible light [1]. Considering that, upconversion nanoparticles are ideal candidate for development of nanosensors in biological applications as they exhibit high photostability, great emission intensity, and most importantly, being exciting by the NIR irradiation.

UCNPs are a type of inorganic nanocrystals comprised of a host matrix to embed the lanthanide dopants, absorbing dopants to harvest NIR light, and emitting dopants to produce the upconversion photoluminescence. The upconversion refers to a non-linear optical process in which the nanoparticle sequentially absorbs two or more low-energy NIR photon and emits a high-energy photon of short wavelength. In contrast with multiphoton process used to excite organic fluorescent dyes or other photoluminescent nanoparticles, the absorption of photons in UCNPs occurs via a real energy level and the upconversion process is several orders of magnitude more efficient, allowing the nanoparticles to be excited at low excitation energy (typically, 1-1000 W/cm^2) [2]. The upconversion emission originates from this energy absorption and transfer process among lanthanide dopants in the host matrix. Therefore, the spectral properties and emission efficiency of UCNPs are largely determined by the selection of UCNP host matrix, the dopant ions, and their doping level. For example, Er^{3+} doped $NaYF_4$:Yb nanocrystals brings about mainly green (Em = 540 nm) and red emission (Em = 655 nm), while Tm^{3+} doped $NaYF_4$: Yb results in dominant blue (Em = 450, 475 nm) and NIR emission (Em = 800 nm) (Fig. 2.1).

The low energy NIR excitation together with the tunable upconversion emission of UCNPs have found them a variety of applications in biomedicine, such as high contrast bioimaging [3, 4], drug delivery [5], photodynamic therapy [4-7], and photothermal therapy [8]. These outstanding properties stemming from the upconversion photoluminescence have also be recognized for enormous potential in biosensing [9, 10]. In this chapter, we present a comprehensive review that covers the main aspects of the development and applications of UCNP-based biosensors, including discussion of optical merits of UCNPs in the context of biosensing, nanoparticle synthesis strategies, design of UCNP nanosensors, and examples of applications of UCNP-based biosensors.

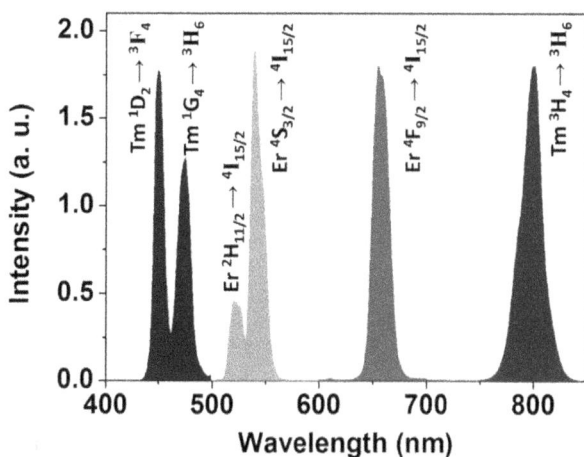

Fig. 2.1. Luminescence emission of UCNPs, blue (Em = 450, 475 nm) and NIR emission (Em = 800 nm) from Tm^{3+} doped nanocrystal, and green (Em = 540 nm) and red emission (Em = 655 nm) from Er^{3+} doped nanocrystal.

2.2. Optical Properties of UCNPs

The upconversion photoluminescence process takes place via a combination of several complex optical pathways, such as ground state absorption, excited state absorption, and energy transfer [11]. In a simplified upconversion process, an incoming NIR photon pumps the absorbing ion (typically Yb^{3+}) from its ground state to an excited state from which a non-radiative energy transferred to the neighboring emitting ion (typically Er^{3+}, Tm^{3+}, and Ho^{3+}) takes place, and promotes it to an intermediate excited state. Another incoming NIR photon stimulates the sequential process of the non-radiative energy transfer, resulting in the same absorbing ion being transferred to the higher excited state. Following the energy transfer, the emitting ions relax to their ground state, while the activator ion at its higher excited state undergoes non-radiative relaxation to the lower energy states, followed by radiative decays with characteristic upconversion emission and eventually returning to its ground state (Fig. 2.2). This unique photoluminescence process ensure UCNPs with many excellent optical properties for biosensing. In this section, a brief description of the optical properties of UCNPs is presented.

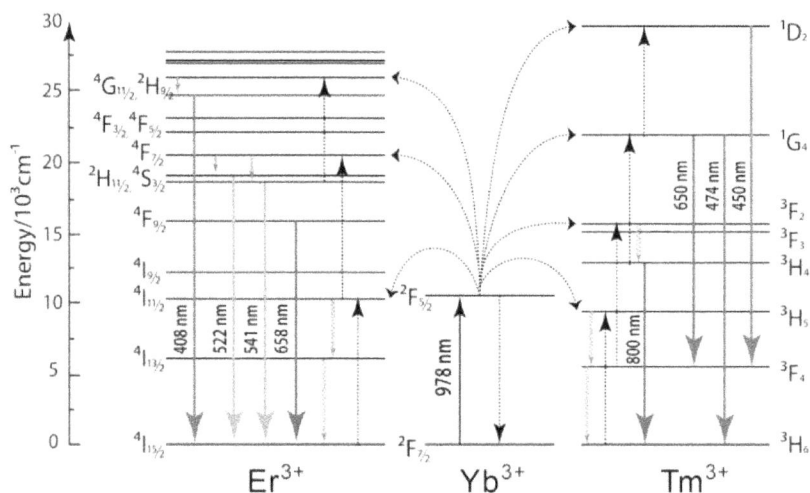

Fig. 2.2. Under NIR laser irradiation, energy transfer mechanisms illustrating upconversion energy transfer processes in Yb^{3+}, Er^{3+}/Tm^{3+} co-doped UCNPs.

2.2.1. Large anti-Stokes Shift and Narrow Multicolour Emission

Generally, fluorescence emitters rely on a single photon excitation occurring in UV or visible spectral range to produce emission with the lower photon energies, the process known as Stokes emission. In contrast, upconversion nanoparticles sequentially absorb two or more low-energy photons at the higher wavelength (*e. g.* 980 nm) producing anti-Stokes emission at the shorter wavelength. This multiphoton process results in a large anti-Stokes shift (up to 500 nm) between the excitation and emission wavelengths, allowing more efficient and easy separation of the photoluminescence and excitation light. Additionally, the emission peaks of UCNPs are narrow-band in comparison with the conventional photoluminescent nanoparticles, with the emission bandwidths being typically ~20 nm. Meanwhile, UCNPs are characterized by a unique set of distinguishable emission peaks. These multicolour peaks are narrow and featured, usually excited simultaneously using a single NIR continuous-wave excitation source. Hence, the multicolour property of UCNPs is beneficial in the design of ratiometric biosensors, where one emission peak is used as a constant reference and a second emission is modulated by the analyte. On the other hand, the multiple emission of UCNPs can also be harnessed for multiplexed detection of different analytes by using the well-separated emissions of UCNPs. This technology allows for simultaneous identification and quantifications of multiple distinctive analytes in complex environment [12]. For example, Rantanen and co-workers fabricated a multiplexed detector for DNA by using the two distinct emission bands of UCNPs (NaYF$_4$:Yb,Er) [13]. In their design, UCNPs were coupled with two different capture oligonucleotides, which would specifically detect the target DNA label with Alexa Fluor 546 or Alex Fluor 700. Upon NIR excitation, the green and red emissions from UCNPs were selectively absorbed by the two fluorophore labels, leading to excited emission of the acceptors at 600 nm and 740 nm, respectively. DNA hybridization between different strands was therefore detected simultaneously without mutual interference [13].

2.2.2. Tuneable Upconversion Luminescence

The capability of manipulating the output colour is another fascinating property of UCNPs, especially when the nanoparticles are used as a donor in the luminescence resonance energy transfer (LRET)-mediated sensing application. LRET is a non-radiative energy transfer process happening between an energy transfer donor and an energy transfer acceptor in a restricted distance (1-10 nm). The on/off energy transfer between UCNP (donor) and fluorophore (acceptor) can be activated or deactivated upon the recognition of the analytes, which provides an effective and simple way for the development of UCNPs-based biosensors. Based on this strategy, a variety of sensing platforms have recently been developed for bioanalysis with improved sensitivity and efficiency [14]. To obtain the high sensitivity in detection, it is critical to choose a good LRET pair with emission of the donor (UCNPs) overlapping well with the absorption band of the acceptor. To this end, UCNPs can act as a flexible donor to provide the diverse selection of emission wavelengths and intensities to the desired receptors. The emission color and intensities of UCNPs is tuneable by various strategies, encompassing varying the host lattices [15], controlling the doping composition [16] and concentration [17], and exploiting the core-shell structure [18], which have been extensively reviewed [19]. These emission profiles of UCNPs also provide opportunities for engineering UCNPs-based optical biosensors, where UCNPs serve as the signaling moiety to report the levels of specific analyte.

2.2.3. Non-invasive and Sensitive Bioassay in Deep Tissues

Capabilities for non-invasive and high-sensitive detection of specific biomolecules in deep tissue is another compelling reason for the development of UCNPs-based nanosensors. In optical biosensing, light travelling through biological tissues undergoes multiple events of absorption and scattering that will contribute to the attenuation of the light. The absorption is mainly caused by biomolecules such as haemoglobin, oxyhaemoglobin in the blood – in UV/visible part of the spectrum; and water – in the NIR. The scattering process takes place on the boundaries between optical interfaces, being more efficient on high refractive-index structures, including lipids, melanin, *etc.* This attenuation effect is minimal when using light at the wavelength range from 700 nm to 1300 nm, a biological tissue transparency window [1]. The characteristic excitation of a particular type of UCNPs, Tm-doped UCNPs at 980 nm and its NIR emission at ~800 nm fall into this region, minimizing the effects of absorption and scattering, are thus considered advantageous for deep-tissue sensing.

Additionally, the autofluorescence from biological samples is greatly eliminated due to the NIR excitation. As a result, a high contrast of UCNPs and low autofluorescence signal can be detected in a crowded background of biological samples. For example, Li's group exploited the ratiometric emission of $NaYF_4$:Yb,Er,Tm at 800 nm to 660 nm to design an efficient and highly sensitive probe for the detection of methylmercury ions ($MeHg^+$) in living animal [20]. In this study, UCNPs were surface modified with a NIR dye heptamethine cyanine dye (hCy7'), which was responsive to $MeHg^+$ and thereby changed its absorption maxima from 670 nm to 845 nm. In presence of $MeHg^+$, energy transfer between hCy7' and UCNPs was increased, resulting in the decrease in NIR emission of UCNPs to be detected *in vivo* by upconversion bioimaging [20].

2.2.4. Superb Photostability

A single fluorescent dye molecule (*e.g.* Rhodamine 6G) typically survives about 1 million excitation-emission cycles, followed by transition to a triplet metastable state from where it reacts with neighboring molecules and undergoes an irreversible transition to the dark state, leading to the cessation of its fluorescence. Fluorescence intermittency or blinking is another consideration for bioimaging *in vivo*. Recent studies on the photostability of a single UCNP showed virtually no photobleaching of the particle during the 1 h of continuous illumination with a 10 mW 980 nm laser [21]. Additionally, no photoblinking of UCNPs at timescales down to 1 ms was observed [21], making this photoluminescence material attractive as optical contrast probe with real-time monitoring and long-term tracking capabilities. For instance, Lai *et al.* developed a UCNPs-based nanohybrid as a drug delivery vehicle as well as a real-time monitoring sensor of drug release within the tested 24 hours in live cells [22]. The multicolor UCNPs was coated with a mesoporous silica to allow the loading of chemotherapeutic drugs in the nanohybrid. As the result of energy transfer from UCNPs to drugs, the emission in UV-vis range was quenched, while the NIR emission was retained. The release of drugs led to the reduction in energy transfer, and thus the emission in UV-vis range was switched on. By virtue of the superb photostability of UCNPs, the drug release from this drug carrier was then monitored by the ratiometric signal of UV-vis and NIR emission of UCNPs [22].

2.3. Development of UCNP-based Biosensors

Upconversion nanoparticle itself is inert to the surrounding microenvironment, while the UCNPs-based nanosensors can be designed by integrating responsive molecules, such as molecular chemosensors with this upconversion nanomaterial. Two different sensing mechanisms have been reported in the applications of UCNPs-based biosensors, namely, inner filter effect and luminescence resonance energy transfer (LRET). In the first case, the emission of UCNPs is quenched due to re-absorption of the emission light from UCNPs by the molecules existing in the detection system. Analytes at different concentration exert a strong or an insignificant filter effect on the luminescence of UCNP, thereby giving an indication of the quantity of analytes [23-28].

LRET is a process in which energy is transferred non-radioactively from an excited donor (UCNP) to an acceptor molecular. LRET requires essentially the emission band of UCNPs overlapping well with the absorption band of the acceptor molecular and the LRET-pair (donor-acceptor) being placed in close proximity to one another (1-10 nm). Therefore, the sensing of responsive nanosensors are determined by the energy transfer efficiency, which are normally modulated either by spectrometric overlapping or changing the distance between the LRET-pair (Fig. 2.3). Fig. 2.3 (a) illustrates the design of UCNPs nanosensor based on the spectrometric overlapping strategy. In the presence of analyte, the absorption spectrum of acceptor is changed, followed by switching on the emission of UCNPs due to the elimination of LRET process. The LRET efficiency is highly dependent on the distance between the donor and acceptor, and therefore UCNPs-based nanosensors can also be designed by manipulating the distance between LRET-pair, which is presented in Fig. 2.3 (b) and (c). On the basis of these two LRET strategies, a myriad of UCNP biosensors have been developed employing the LRET mechanisms to probe a variety of delicate biological processes [14].

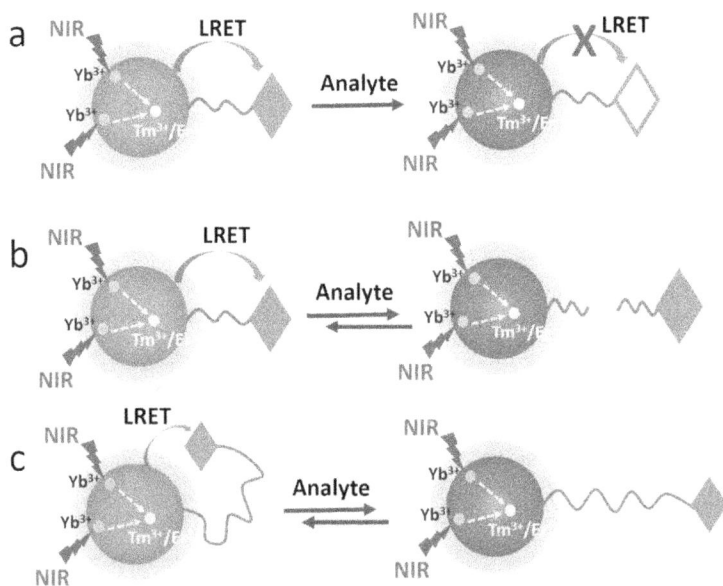

Fig. 2.3. Schematic diagram of the principles of LRET-based UCNPs nanosensors for the detection of analytes. (a) Spectra overlapping between LRET-pair; (b), (c) distance variation between LRET-pair.

In both sensing design approaches (inner filter effect and LRET), it is prerequisite to obtain UCNPs with high luminescence efficiency in order to achieve highly sensitive implementation of UCNP nanosensors. The fabrication of UCNP-donor nanohybrid is also critical to manipulate LRET process for the detection of different analytes. In the following section, the preparation approaches for high-quality UCNPs are summarized, followed by the fabrication strategies for development of UCNP sensing platforms.

2.3.1. Synthesis of UCNPs

Among various types of host materials, lanthanide (Ln) fluorides, such as LnF_3, LnOF, and $MLnF_n$ (M = Li, Na, K or Ba; n = 4 or 5) are regarded as ideal host matrices to produce efficient upconversion nanocrystals. Considering that, a number of synthesis methods have been developed to prepare lanthanide fluoride-based UCNPs with controlled size, shape, crystalline phase, and composition that presents desirable physiochemical properties for their potential applications. The most commonly used methods for the synthesis of UCNPs are thermal decomposition, hydro(solvo)thermal synthesis, and coprecipitation, which are described below.

2.3.1.1. Thermal Decomposition

Thermal decomposition is a well-established method for the synthesis of monodispersed UCNPs with uniformed shape, tailored size and single crystal structure. This strategy is based on the decomposition of organometallic precursors in the presence of organic solvents (e.g. 1-octadecane, ODE) and surfactants (*e. g.* oleic acid, OA, and oleylamine, OM). The commonly used precursors include metallic or lanthanide trifluoroacetate, lanthanide oleates, lanthanide acetates, and lanthanide chlorides. The surfactants usually contain a functional group, including -COOH, and $-NH_2$, to cap the surface of UCNPs for controlling their growth and a long hydrocarbon chain to assist their dispersion in organic solvents.

In general, the synthetic process is conducted at elevated temperature (250-330 °C) in an oxygen-free and anhydrous environment, wherein the precursors decompose to form the nucleus for a particle to grow on. This method was introduced by Yan and co-workers on the preparation of triangular LaF_3 nanoplates [29], and was later improved as a broadly applicable route to produce UCNPs of high quality and narrow size distribution. Various sizes and shapes of UCNPs have been produced by tailoring the experimental parameters, including the reaction temperature, reaction time, nature and concentration of solvents, and the concentration of reagents. Another refined approach was reported by Li and co-worker for the preparation of β-NaYF$_4$:Yb,Er/Tm UCNPs [30]. Their method was demonstrated to be user-friendly in minimizing the use of fluoride reactions and decreasing the amount of toxic by-products generated at high temperature [30].

2.3.1.2. Hydro(solvo)thermal Synthesis

Hydro(solvo)thermal method is performed with the assistance of high temperature and pressure to dissolve solid reactants as well as to speed up the reaction. The possible advantages of this technique are the relatively lower reaction temperature, high-quality crystalline phase of the obtained nanoparticles, and excellent control over the particle size and shape. The main disadvantages are the adoption of specialized reaction vessels (Teflon-lined autoclave) and inability to monitor the particle growth. In a typical process, lanthanide precursors (such as lanthanide nitrites, chlorides, and oxides) and fluoride precursors (such as HF, NH$_4$F, NH$_4$HF$_2$, NaF, and KF) and surfactants are mixed in aqueous solution and placed in an autoclave, then sealed and heated at a temperature between 160 °C and 220 °C. Morphologies of the UCNP product can easily be tuned by varying the reactant concentration, reaction temperature, reaction time, and pH of the solution. This method was firstly reported by Li's group on the synthesis of NaYF$_4$, YF$_3$, LaF$_3$ and YbF$_3$ nanocrystals [31]. Another example of the hydro(solvo)thermal synthesis was reported by Zhao and co-workers to generate monodispersed β-NaYF$_4$ with various morphologies, such as nanorods, nanotubes, and flower-patterned nanodisks [32].

2.3.1.3. Coprecipitation

Coprecipitation approach is the most convenient and simplest way to prepare NaYF$_4$ UCNPs, since no costly equipment, complex procedures, and stringent reaction conditions are required for this synthesis. In a typical example, ethylenediaminetetraacetic acid (EDTA) was used as a chelate agent to form a lanthanide-EDTA complex, followed by a rapid injection of this complex to a vigorously stirred NaF solution [33]. This process was helpful in forming a homogenous nucleus for subsequent growth of nanoparticles. The particle size can be effectively controlled from 37 nm to 166 nm by adjusting the molar ratio of EDTA to lanthanide salts [33]. Normally, α-NaYF$_4$ UCNPs obtained using this method display low photoluminescent yield. In view of that, a post-treatment by annealing is required to drive transition of the particles from cubic to hexagonal phase, which results in the brighter UCNPs [33]. Haase and co-workers have demonstrated successful production of water-dispersible β-NaYF$_4$ UCNPs without the need of the calcination step [2]. Besides the use of EDTA, other surface ligands, such as polyethylenimine (PEI) [32] and polyvinylpyrrolidone [34] were also employed to control the nanoparticle growth and yielded nanoparticles capped with these polymers.

2.3.2. Fabrication of UCNPs-based Sensing Platforms

The hydrophilicity and stable dispersity in biological conditions are the prerequisite for nanoparticles to be used in bioassays. However, UCNPs prepared by the methods described above are generally hydrophobic owing to the hydrophobic nature of the capping reagents (*e.g.* OA or OM), which hampers their application in biosensing. In order to transfer these hydrophobic UCNPs into hydrophilic, a number of surface modification methods have been developed, including ligand exchange [35, 36], ligand oxidation [37-39], ligand removal [40, 41], ligand attraction [42, 43], layer-by-layer assembly [44, 45], and surface silanization [46]. These surface engineering methods not only render UCNPs water solubility, but also endow UCNPs with reactive groups for subsequent conjugation to responsive molecules for development the UCNP-based nanosensors [47-49].

A successful implementation of UCNP nanosensors heavily relies on the responsive recognition molecules. Therefore, the integration of recognition components with the UCNP-sensing system is crucial for the development of UCNPs-based biosensors. For the UCNP sensors in forms of films or beads, UCNPs were mixed with detection molecules in a film matrix (*e. g.* polystyrene), where the detection probe molecules exert an effect on upconverted emission of UCNPs in response to analytes [25, 26]. In this case, LRET between UCNP and responsive probe molecules is unlikely to occur to a substantial extent due to the fact that the distance between donor and acceptor are far above the critical LRET (typically < 10 nm). To fabricate UCNP nanosensors based on LRET, efforts were made to couple recognition agents with UCNP within the LRET distance. To date, the established methods for combining UCNPs with recognition agents include physical adsorption, coordination bond, covalent conjugation, and silica encapsulation (Fig. 2.4).

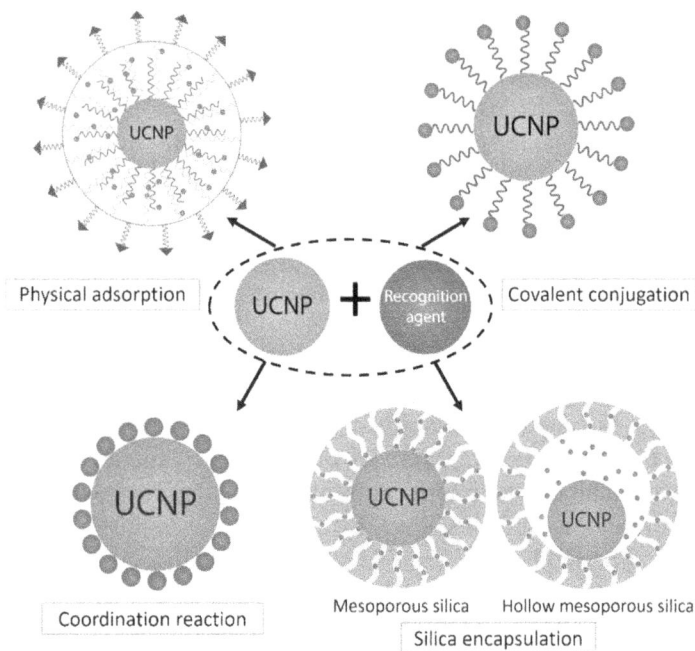

Fig. 2.4. Schematic illustrations of the four strategies for integrating UCNP with recognition agents.

2.3.2.1. Physical Adsorption

In physical adsorption strategy, UCNPs and recognition agents are generally assembled via the hydrophobic interaction or electrostatic attraction. UCNPs prepared by the aforementioned methods are normally capped with a layer of OA, which can facilitate the direct binding of some hydrophobic molecules to UCNPs. For example, Zhang et al. reported the attachment of Cu^{2+} fluorescence sensing molecules, RB-hydrazine, to UCNP surface via the hydrophobic interaction with the OA layer [50]. The hydrophobic arms of some amphiphilic polymers are normally attached to UCNP surface via hydrophobic interaction with UCNP surface ligands, creating a hydrophobic network for storage of hydrophobic recognition agents. For example, Yao and co-workers described the loading of a CN⁻ reactive iridium (III) complex to amphiphilic block polymer coated UCNP, wherein the iridium complex was accommodated in a hydrophobic layer created by the interaction between hydrophobic units of the polymer and oleic acid [51]. In another report, Ding and co-workers modified UCNP with a cyclic oligosaccharide macromolecule γ-cyclodextrin (CD), which served as a doughnut-shaped cavity allowing for loading of both oleate ligands of UCNP and sensing molecules of Rhodamine B derivative [52]. On the other hand, physical adsorption can also make use of the electrostatic interaction between species with opposite charges to deposit recognition molecules on a charged surface of UCNP. For example, negatively charged graphene oxide (GO) nanosheets can be loaded on to positive charged PEI-coated UCNPs [53]. In analogy, positively charged Zn^{2+}-responsive compound was assembled onto polyacrylic acid (PAA)-modified UCNPs (negatively charged) through electrostatic attraction [54].

2.3.2.2. Coordination Bond

Because of the strong coordination capability of abundant lanthanide ions on the UCNPs surface, recognition molecules with functional groups, such as -COOH, -NH₂, and -OH, can be directly immobilized on UCNPs surface via coordination interaction. The coordination is conducted by replacing the original surface capping ligands with the new binding molecules that possess stronger coordination ability toward lanthanide ions. Usually, the ligand exchange reaction involves one or two steps. In on-step method, UCNPs are mixed with excess of new ligands under ultrasonication [55-57]. In two-step processes, surface ligands of UCNP are firstly removed by protonation with HCl at pH 4 [41] or treatment with nitrosonium tetrafluoroborate [36] or tetramethylammonium hydroxide [58], yielding ligand-free UCNP for subsequent attachment of new capping ligands. For example, Liu's group reported the removal of OA on UCNP with substantial amount of acidic ethanol (pH = 1), followed by tagging a Ca^{2+}-sensitive receptor on UCNP to generate a LRET-based nanosensor for Ca^{2+} detection [59].

2.3.2.3. Covalent Coupling

Compared to the physical methods, covalent conjugation between recognition molecules and UCNPs is sufficiently strong to secure the detection reagents remaining in the nanoplatform during application in biological systems. Also, such robust binding is important for the development of UCNP-LRET based sensors, when detachment of recognition molecules is undesirable and will result in a loss of sensing efficiency. In this method, the attachment of recognition molecules to UCNP is generally achieved by cross-linking between carboxylic and amine groups on these two entities. For example, Kumar and Zhang reported the covalent conjugation of amine-containing DNA strand to carboxylic group functionalized UCNP via the carbodiimide coupling reaction [60]. This method presented a versatile strategy for binding amine-containing DNA or aptamers to UCNPs, which are also exploited by other groups [61-63]. Alternatively, recognition molecules with silanol groups can hydrolyze on the surface of silica-coated UCNPs to obtain the covalently linked UCNP-sensing complex [64].

2.3.2.4. Silica Encapsulation

Silica-coating is one of the most popular methods to endow UCNP with aqueous solubility and stability in physiological environments. In an attempt to use the silica-UCNP as delivery vehicles, efforts have been devoted to grow a mesoporous silica ($mSiO_2$) shell or hollow mesoporous silica ($hmSiO_2$) on UCNP. The silica shell on the outer layer of the particle can offer a large storage space for loading recognition agents in the mesoporous channels. A number of biosensors have been developed by incorporating recognition molecules in UCNP@$mSiO_2$ for the detection of hydrogen sulfide [65], Cu^{2+} [66], and pH [67]. A cavity between UCNP and mesoporous silica layer in UCNP@$hmSiO_2$ further advances the design in enabling much higher loading level of recognition agents [68, 69], as compared with UCNP@$mSiO_2$. In addition, the surface of silica is easily conjugated with functional groups, such as amine, thiols, and carboxyl groups, which in turn facilitate the conjugation of biomolecules or targeting ligands to UCNPs.

2.4. Applications of UCNPs in Biosensing

As a novel class of luminophore, UCNPs have been demonstrated its feasibility in the development of multiple *in vitro*, *ex vivo*, and *in vivo* biosensing models as the energy donor [9, 14]. The following section summarized some of the recent studies on development and applications of UCNPs-based biosensors, broadly grouped into sensing of ions, biomolecules, and gas molecules.

2.4.1. UCNPs Nanosensors for the Detection of Ions

Hazardous ions, such as CN⁻ and Hg^+ are known as extremely toxic chemicals that can cause serious problems to the environment and health. Selective detection and bioimaging of these ions are of great importance for the biological systems. On the other hand, metal ions such as Fe^{3+}, Cu^{2+}, Zn^{2+}, Na^+, K^+, and Ca^{2+} are essential trace elements involved in many biological activities. Imbalance of these metal ions are implicated in various disorders for life processes [27, 28, 50, 52, 54]. Monitoring the level of these ions is thus significant for physiological concerns.

In this aspect, by exploiting the LRET-based mechanism, the prepared UCNP nanosensors provide particular useful tools for highly sensitive and selective detection of these ions *in vitro* and *in vivo*. Here, the ions responsive molecules are typically integrated with UCNPs in a nanocomplex, where the upconversion emission of UCNPs dominates the energy transfer efficiency from UCNPs to responsive molecules. This energy transfer can also be manipulated by a specific recognition process between responsive molecules and analyte [10]. The inhibition and promotion of energy transfer from UCNP to indicating probes will lead to changes in luminescent intensity of UCNP, thereby providing the evaluation of concentrations of target ions [70].

In a typical example, Li's group reported a UCNPs-based ratiometric luminescence probe for sensing of CN⁻. The responsive chromophoric iridium (III) complex was employed for manipulating the green and red emissions of UCNPs [71]. As the result of specific reaction between iridium (III) complex and CN⁻, the energy transfer from UCNPs to iridium (III) complex was diminished, followed by the recovery of green emission of UCNPs (Fig. 2.5) [71]. This UCNPs nanosensor exhibited low detection limit of 0.18 μM for CN⁻ in the aqueous solutions, however was not applicable in pure water due to hydrophobic nature of the iridium (III) complex [71]. Further study conducted by the same group showed that the water solubility was significantly improved when coating the UCNPs surface with amphiphilic block-polymers. The CN⁻-reactive iridium (III) complex, (ppy)₂Ir(dmpp)]PF₆, was loaded to the formed hydrophobic layer for the detection of CN⁻ in pure water [51].

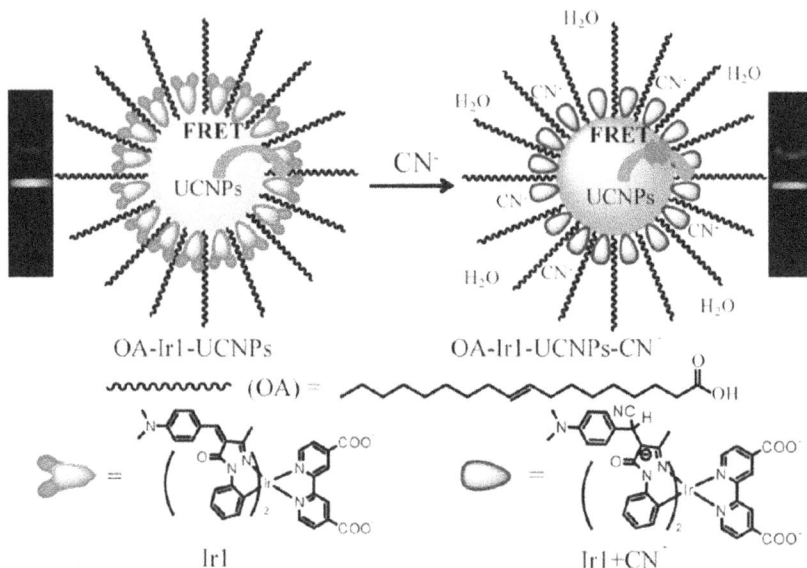

Fig. 2.5. Recognition mechanism and the LRET process of Ir1-UCNPs towards CN⁻. Reprinted with permission from ref. [71]. *Copyright 2011 American Chemical Society.*

In another work, Li's group fabricated a Hg^{2+} sensor by assembling a water soluble Hg^{2+} probe ruthernium complex N719 directly to UCNP ($NaYF_4$:Yb,Er,Tm) via ligand exchange approach (Fig. 2.6) [55]. Upon interaction with Hg^{2+}, maximum absorption of N719 shifted from 541 to 485 nm, resulting in the decrease in spectral overlap between green emission of UCNPs and the absorption of N719. Therefore, the green emission of UCNPs was recovered, and the Hg^{2+} levels was determined [55]. This designed N719-UCNP nanosensor has also shown to be applicable in monitoring the distribution of Hg^{2+} in living cells by upconversion luminescence imaging [55].

2.4.2. UCNPs Nanosensors for the Detection of Biomolecules

Molecular recognition and specific interaction between biotin and avidin, antibody and antigen, aptamer and receptor, and molecular binding via enzyme reaction and DNA hybridization have provided reliable and versatile routes for the fabrication of biosensors with sensitivity and specificity. These strong interactions can also be utilized to manipulate the interaction between UCNPs and analytes for detection of molecular analytes at the nanoscale level by triggering or inhibiting the energy transfer process.

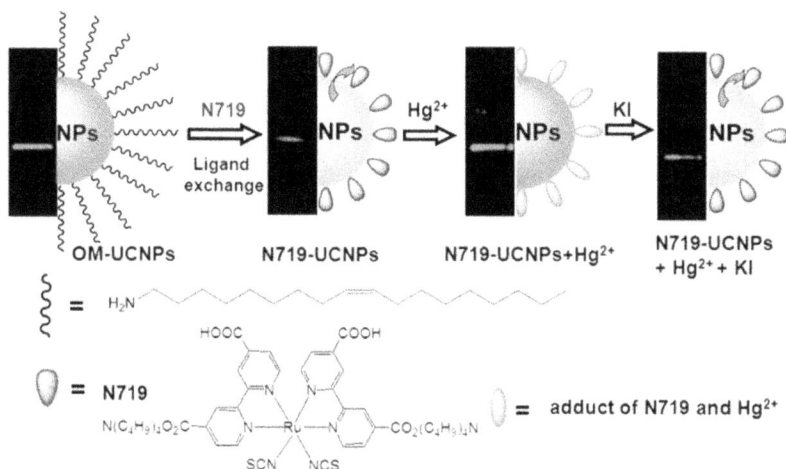

Fig. 2.6. Schematic illustration of preparation of N719-modified NaYF$_4$:Yb,Er,Tm and its sensing functionality. Reprinted with permission from ref. [55]. *Copyright 2011 American Chemical Society.*

2.4.2.1. Application of UCNPs in DNA Detection

DNA hybridization, the complementary interaction between probe DNA and sample DNA, is a major technique for the detection of genetic disease sequences. The DNA hybridization can be monitored by developing UCNPs-based sensors, where the sample DNA can be detected through ratiometric changes in luminescence emission of UCNPs. Based on this approach, Zhang and co-workers reported a simple designed UCNPs nanosensors for sensing of DNA [72]. This UCNP nanosensor was developed by conjugating capture oligonucleotide to green-emitting UCNPs, and labeling the reporter oligonucleotide with green-absorbing fluorescent dye TAMRA (Fig. 2.7 (a)) [72]. In presence of target oligonucleotide, the hybridization occurred, leading to the upconverted green emission being quenched by energy transfer between UCNP and TAMRA [72].

In an alternative way, Zhang and co-workers presented the use of an intercalating dye SYBR Green I as a fluorescent indicator to bind to the hybridized double strands of DNA [61]. After hybridization, SYBR

Green I was intercalated between target nucleotide and UCNP conjugated probe nucleotide (Fig. 2.7 (b)) [61]. Under NIR illumination, the green emission from UCNP was transferred to SYBR Green I, therefore, the DNA detection can be achieved by recording the enhancement in emission of SYBR Green I and decreased luminescence of UCNP [61]. These two schemes have also be applied to the LRET detections of DNA in other studies [13, 73, 74].

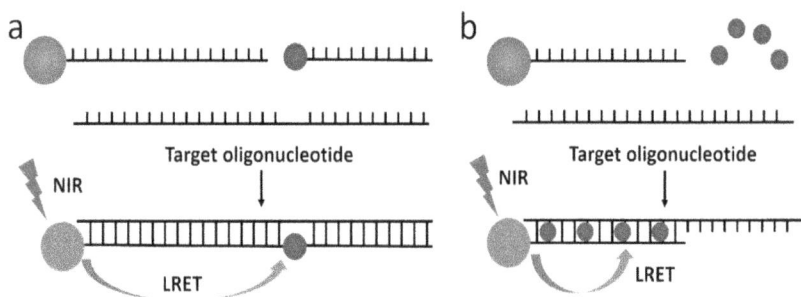

Fig. 2.7. Schematic illustration of design of UCNP-LRET oligonucleotide sensors based on the hybridization of DNA.

In design of LRET based UCNP sensors, the hybridization process of DNA can also be used together with the special interaction between graphene oxide (GO) and DNA. Single-stranded DNA (ssDNA) can be firmly attached to GO surface via the nucleobases-graphene binding, while double-stranded DNA (dsDNA) showed weak attachment towards GO. The changes on the binding affinity between ssDNA and dsDNA to GO are therefore adopted to discriminate DNA sequences. Kanaras, et al. exploited the LRET between UCNPs and GO and the hybridization of DNA to prepare a sensitive DNA sensor [62]. UCNPs were coated with a silica layer of 11 nm, and further attached with ssDNA via covalent conjugation (Fig. 2.8). When the nanocomplex interacted with GO templates, the spectral overlap between fluorescence emission of UCNP and the absorption of GO resulted in quenching the emission of UCNPs (Fig. 2.8). However, in the presence of the complementary DNA strands, the dsDNAs formed by hybridization, leading to the detachment of UCNP from GO and therefore the reduced quenching effect of GO to UCNP. The detection limit of this new method was shown to be 5 pM, demonstrating the high sensitivity of the developed sensors [62].

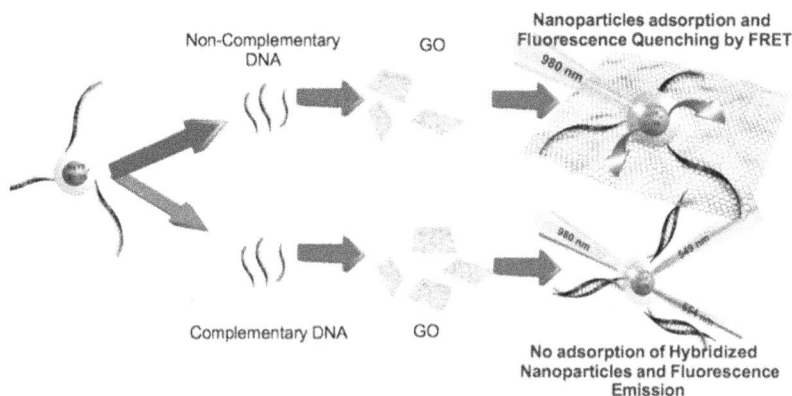

Fig. 2.8. Schematic illustration of the DNA sensor based on DNA hybridization and LRET between UCNP and GO. Reprinted with permission from ref. [62]. *Copyright 2015 American Chemical Society.*

2.4.2.2. Application of UCNPs in the Detection of Small Molecules, Proteins and Enzymes

Determination the levels of biomolecules, such as amino acids, proteins, and enzymes in biological samples plays a key role in understanding their physiological function in live systems, though remains challenging. Rapid development of biosensing technology makes detection of these molecules *in situ* possible. Recent studies on engineering of UCNPs-based approaches accelerates the applications of optical biosensors for the background-free detection of these molecules with high sensitivity and selectivity. Up to now, a number of UCNP sensors have been developed to detect a wide range of biomolecules, such as thrombin [75, 76], antigens [77], hemoglobin [78], adenosine triphosphate [79], cysteine/homocysteine [69], glucose [80], glutathione [81], matrix metalloproteinase [82], mycotoxins [83], siRNA [84], miroRNA [85], messenger RNA [86]. These representative studies provide paradigms for the design of UCNP-sensors in bioanalysis applications. In this section, we will present some examples of UCNPs-based optical biosensors for the quantification of proteins and enzymes.

By using UCNPs as the energy donor, and gold nanoparticle as the energy acceptor, Li and co-workers fabricated an effective LRET platform for the detection of trace amounts of avidin [44]. In their work, both UCNPs (NaYF$_4$:Yb,Er) and gold nanoparticles were conjugated with biotin, and the quantitative analysis of avidin was achieved when

UCNP and gold NPs were bridged by avidin, and luminescence of UCNPs were gradually quenched by gold NPs with increasing amounts of avidin [44]. The strategy of energy transfer from UCNPs to gold nanoparticles has also been exploited for the detection of goat anti-human IgG antibodies by Xu's group [87]. In their study, UCNP and gold NPs were conjugated to rabbit anti-goat IgG antibodies and human IgG antibodies, respectively (Fig. 2.9). Upon the addition of goat anti-human antibodies, UCNP and gold nanoparticles were linked via the sandwich-type immunoreaction, which in turn initiated the LRET between UCNP and gold NPs , thus quenching the emission from UCNP [87].

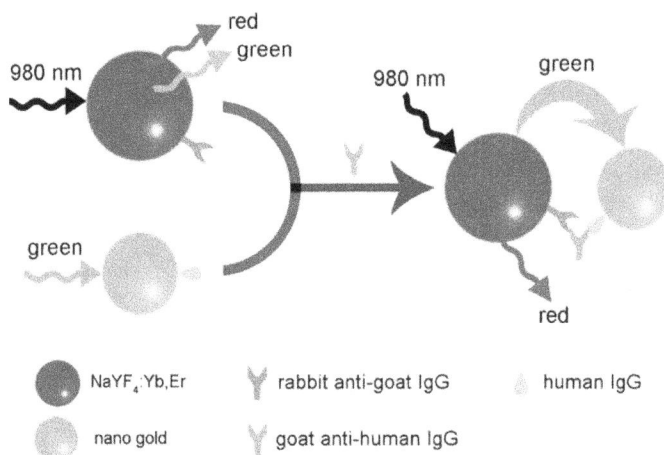

Fig. 2.9. Schematic illustration of detecting anti-human IgG antibodies using the LRET process between UCNP and gold NPs. Reprinted with permission from ref. [87]. *Copyright 2009 American Chemical Society.*

Besides graphene oxide and gold NPs, carbon nanoparticles with strong quenching ability to fluorescence have also been used to construct UCNP-based LRET sensors. The group of Pang reported the development of an aptamer biosensor for the detection of thrombin based on the LRET between UCNPs to carbon nanoparticles [79]. In their work, a thrombin aptamer was covalently conjugated to PAA modified UCNP, allowing the further binding of UCNPs to carbon NPs via π-π stacking interaction [79]. As a result, the fluorescence of UCNPs was quenched when the two NPs were brought into close proximity. In the presence of thrombin, the luminescence from UCNPs was restored due to the released binding of carbon NPs to UCNPs, induced by the thrombin-mediated quadruplex structure formation of aptamer. Because of the strong fluorescence quenching ability and excellent biocompatibility of carbon NPs, the formed UCNP-carbon NPs biosensor demonstrated favorable sensing performance in human plasma samples [79].

2.4.3. UCNPs Nanosensors for the Detection of Gas Molecules

The determination of gases contents of oxygen, carbon dioxide, ammonia, hydrogen sulfide, and many other gases is important to correlate their roles with diseases in living systems, which in turn provide crucial information to understand their physiological and pathological functions and to develop advanced medicine for disease treatment. In this regard, the optical sensing using UCNP is particularly useful in affording the photostable upconverted luminescence in biological samples.

The first example of UCNP-based gas sensor was reported by the group of Wolfbeis. The sensor was prepared by embedding the pH sensitive probe bromothymol blue (BTB) with $NaYF_4$:Yb,Er in CO_2 permeable polystyrene film [24]. The absorption of BTB overlapped with the green and red emission of UCNP. Once CO_2 penetrated the film, the pH of the microenvironment was changed, and the absorption of BTB was decreased due to protonation, followed by the enhancement of luminescence from UCNPs. The principle of pH changes can also be applied for the development of UCNPs-based ammonia sensors [88, 89]. These optical sensors for ammonia are based on a change in either color or fluorescence of the indicator dye from transduction of pH.

It has been well-known that the transition metal complexes, such as ruthenium (II) and iridium (III) complexes are oxygen-sensitive, and can be used as the sensor for the detection of oxygen levels. More importantly, the absorption spectra of these metal complexes are well matched with the blue or green emissions of UCNPs. Taking advantages of these properties, Su's et al. presented the first integration of UCNPs with oxygen-sensing ruthenium (II) complex into one nanoplatform for oxygen sensing [64]. In a subsequent study, Wolfbeis and co-workers demonstrated the oxygen detection by using UCNPs as the nanolamp, and an iridium (III) complex as the switchable molecular probe for oxygen [26]. The absorbance of iridium complex strongly overlapped with the blue emission of UCNPs. Therefore, the NIR-excited blue emission of UCNPs was demonstrated to be absorbed by iridium complex to generate its green-yellow luminescence, which was quenched during the sensing of oxygen [26].

Recently, Shi's group demonstrated the application of UCNPs-[Ru(dpp)$_3$]$^{2+}$Cl$_2$ systems for the diagnosis of hypoxic level in living organisms by exploiting the UCNPs-based LRET approach [68]. The nanosensor was prepared by loading oxygen-sensitive ruthenium (II) complex into the hollow cavity of UCNP@hmSiO$_2$, wherein the distance between UCNP donor and ruthenium acceptor was maintained to be within 10 nm (Fig. 2.10). Under NIR exposure, the energy transfer between UCNP and ruthenium (II) complex resulted in the illumination of ruthenium (II). Since the luminescence intensity of ruthenium (II) complex is oxygen dependent, the hyperoxic or hypoxic conditions in both cells and live zebrafishes were then determined by spectrometric analysis and bioimaging.

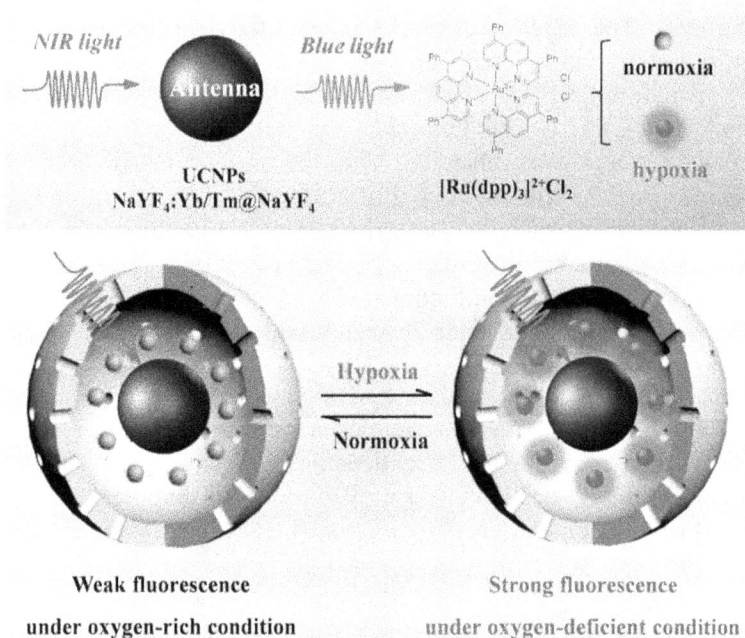

Fig. 2.10. Schematic illustration of UCNP-based oxygen nanosensor by impregnating the oxygen-responsive indicator into hollow cavity of UCNP@hmSiO$_2$. Reprinted with permission from ref. [68]. *Copyright 2014 American Chemical Society.*

Based on the similar LRET strategy, Liu and co-workers prepared a UCNPs nanosensor for the highly sensitive and selective detection of HS⁻ in living cells [65]. The UCNP sensing platform was designed by coating a mesoporous silica shell on UCNP, and immobilizing the HS⁻ sensitive probe merocyanine (MC) into the mesopores of the mSiO$_2$ layer. Upon the interaction of MC with HS⁻ to form MCSH, the absorption at ~548 nm was decreased. Therefore, the green luminescence emission from UCNPs was increased as the diminishment of energy transfer from UCNPs to dye molecules. In addition, ratiometric luminescence detection of HS⁻ was achieved by using the emission at 800 nm as the internal standard.

2.5. Conclusions

Research exploring the use of luminescent nanoparticles for optical sensing is driven by the increasing demands for high accuracy and sensitivity in bioassays. In this sense, the emerging upconversion nanoparticles that are illuminated by NIR light provide numerous opportunities for sensitive biological sensing, particular for the detection of biomolecules in complicated biological conditions. UCNP demonstrate remarkable optical properties, including large anti-Stokes shift, narrow emission bands, tuneable emission spectrum, and excellent photostability. These merits of UCNPs together with the flexible surface chemistry and low toxicity have made UCNPs a successful luminescent platform for the development of biosensors for the detection of ions, biomolecules, and biological gases that are summarized in this chapter.

Although great potentials of UCNPs have been widely recognized, challenges still exist in implementation of UCNPs in biosensing. First of all, small and bright UCNPs are required in the context of biological optical imaging and sensing. However, the reproducible preparation of small UCNPs with improved upconversion luminescence efficiency remains challenging. Secondly, significant quenching of upconversion emission in water solution is undesirable for their application in biological systems. Thirdly, the elaborate design of LRET UCNP sensors entails further investigation to optimize the energy transfer between UCNP and sensing probe, and to elucidate the contribution of LRET and emission-reabsorption to the sensing efficiency. Fourthly, besides the LRET, novel sensing mechanism for the design of UCNPs-based nanosensors are demanded. Lastly, applications of the designed UCNPs in addressing the practical medical problems needs to be explored. Since this nanotechnology is progressing gradually, we can envision that more applications of UCNPs-based biosensing will be achieved in the not-too-distant future.

Acknowledgements

We wish to acknowledge the financial supports from Australian Research Council (DE170100092, DP170104749), National Health and Medical Research Council (APP1125794), and National Natural Science Foundation of China (21601076), and Russian Federation (Megagrant: 14.Z50.31.0022). This work is also partially supported by the Australian Research Council through its Centre of Excellence scheme (CE140100003).

References

[1]. P. N. Prasad, Introduction to biophotonics, *John Wiley & Sons*, 2004.

[2]. S. Heer, K. Kömpe, H. U. Güdel, M. Haase, Highly efficient multicolour upconversion emission in transparent colloids of lanthanide-doped NaYF$_4$ nanocrystals, *Advanced Materials*, 16, 2004, pp. 2102-2105.

[3]. A. Nadort, V. K. Sreenivasan, Z. Song, E. A. Grebenik, A. V. Nechaev, V. A. Semchishen, V. Y. Panchenko, A. V. Zvyagin, Quantitative imaging of single upconversion nanoparticles in biological tissue, *PloS One*, 8, 2013, e63292.

[4]. L. Liang, Y. Lu, R. Zhang, A. Care, T. A. Ortega, S. M. Deyev, Y. Qian, A. V. Zvyagin, Upconversion nanoparticles mediated deep-penetrating photodynamic therapy of KillerRed, *Acta Biomaterialia,* 51, 2017, pp. 461-470.

[5]. D. Yang, Z. Hou, Z. Cheng, C. Li, J. Lin, Current advances in lanthanide ion (Ln^{3+})-based upconversion nanomaterials for drug delivery, *Chemical Society Reviews,* 44, 2015, pp. 1416-1448.

[6]. L. Liang, A. Care, R. Zhang, Y. Lu, N. H. Packer, A. Sunna, Y. Qian, A. V. Zvyagin, Facile assembly of functional upconversion nanoparticles for targeted cancer imaging and photodynamic therapy, *ACS Applied Materials & Interfaces,* 8, 2016, pp. 11945-11953.

[7]. E. Khaydukov, K. Mironova, V. Semchishen, A. Generalova, A. Nechaev, D. Khochenkov, E. Stepanova, O. Lebedev, A. Zvyagin, S. Deyev, Riboflavin photoactivation by upconversion nanoparticles for cancer treatment, *Scientific Reports,* 2016.

[8]. L. Cheng, K. Yang, Y. Li, X. Zeng, M. Shao, S.-T. Lee, Z. Liu, Multifunctional nanoparticles for upconversion luminescence/MR

multimodal imaging and magnetically targeted photothermal therapy, *Biomaterials*, 33, 2012, pp. 2215-2222.

[9]. D. E. Achatz, R. Ali, O. S. Wolfbeis, Luminescent chemical sensing, biosensing, and screening using upconverting nanoparticles, Luminescence applied in sensor science, *Springer*, 2010, pp. 29-50.

[10]. S. Hao, G. Chen, C. Yang, Sensing using rare-earth-doped upconversion nanoparticles, *Theranostics*, 3, 2013, pp. 331-345.

[11]. A. Nadort, J. Zhao, E. M. Goldys, Lanthanide upconversion luminescence at the nanoscale: fundamentals and optical properties, *Nanoscale*, 8, 2016, pp. 13099-13130.

[12]. S. Wu, N. Duan, Z. Shi, C. Fang, Z. Wang, Simultaneous aptasensor for multiplex pathogenic bacteria detection based on multicolor upconversion nanoparticles labels, *Analytical Chemistry*, 86, 2014, pp. 3100-3107.

[13]. T. Rantanen, M. -L. Järvenpää, J. Vuojola, R. Arppe, K. Kuningas, T. Soukka, Upconverting phosphors in a dual-parameter LRET-based hybridization assay, *Analyst*, 134, 2009, pp. 1713-1716.

[14]. Q. Su, W. Feng, D. Yang, F. Li, Resonance Energy Transfer in Upconversion Nanoplatforms for Selective Biodetection, *Accounts of Chemical Research*, 50, 2017, pp. 32-40.

[15]. J. Zhao, Z. Lu, Y. Yin, C. McRae, J. A. Piper, J. M. Dawes, D. Jin, E. M. Goldys, Upconversion luminescence with tunable lifetime in $NaYF_4$: Yb, Er nanocrystals: role of nanocrystal size, *Nanoscale*, 5, 2013, pp. 944-952.

[16]. F. Wang, X. Liu, Upconversion multicolor fine-tuning: visible to near-infrared emission from lanthanide-doped $NaYF_4$ nanoparticles, *Journal of the American Chemical Society*, 130, 2008, pp. 5642-5643.

[17]. Y. Lu, J. Zhao, R. Zhang, Y. Liu, D. Liu, E. M. Goldys, X. Yang, P. Xi, A. Sunna, J. Lu, Tunable lifetime multiplexing using luminescent nanocrystals, *Nature Photonics*, 8, 2014, pp. 32-36.

[18]. F. Wang, R. Deng, J. Wang, Q. Wang, Y. Han, H. Zhu, X. Chen, X. Liu, Tuning upconversion through energy migration in core-shell nanoparticles, *Nature Materials*, 10, 2011, pp. 968-973.

[19]. H. H. Gorris, O. S. Wolfbeis, Photon-upconverting nanoparticles for optical encoding and multiplexing of cells, biomolecules, and microspheres, *Angewandte Chemie International Edition*, 52, 2013, pp. 3584-3600.

[20]. Y. Liu, M. Chen, T. Cao, Y. Sun, C. Li, Q. Liu, T. Yang, L. Yao, W. Feng, F. Li, A cyanine-modified nanosystem for in vivo upconversion luminescence bioimaging of methylmercury, *Journal of the American Chemical Society*, 135, 2013, pp. 9869-9876.

[21]. S. Wu, G. Han, D. J. Milliron, S. Aloni, V. Altoe, D. V. Talapin, B. E. Cohen, P. J. Schuck, Non-blinking and photostable upconverted luminescence from single lanthanide-doped nanocrystals, *Proceedings of the National Academy of Sciences*, 106, 2009, pp. 10917-10921.

[22]. J. Lai, B. P. Shah, Y. Zhang, L. Yang, K. -B. Lee, Real-time monitoring of ATP-responsive drug release using mesoporous-silica-coated multicolor upconversion nanoparticles, *ACS Nano*, 9, 2015, pp. 5234-5245.

[23]. H. Chen, J. Ren, Sensitive determination of chromium (VI) based on the inner filter effect of upconversion luminescent nanoparticles (NaYF$_4$: Yb^{3+}, Er^{3+}), *Talanta*, 99, 2012, pp. 404-408.

[24]. R. Ali, S. M. Saleh, R. J. Meier, H. A. Azab, I. I. Abdelgawad, O. S. Wolfbeis, Upconverting nanoparticle based optical sensor for carbon dioxide, *Sensors and Actuators B: Chemical*, 150, 2010, pp. 126-131.

[25]. R. J. Meier, J. M. Simbürger, T. Soukka, M. Schäferling, Background-free referenced luminescence sensing and imaging of pH using upconverting phosphors and color camera read-out, *Analytical Chemistry*, 86, 2014, pp. 5535-5540.

[26]. D. E. Achatz, R. J. Meier, L. H. Fischer, O. S. Wolfbeis, Luminescent sensing of oxygen using a quenchable probe and upconverting nanoparticles, *Angewandte Chemie International Edition*, 50, 2011, pp. 260-263.

[27]. L. Xie, Y. Qin, H.-Y. Chen, Polymeric optodes based on upconverting nanorods for fluorescent measurements of pH and metal ions in blood samples, *Analytical Chemistry*, 84, 2012, pp. 1969-1974.

[28]. L. Xie, Y. Qin, H. -Y. Chen, Direct fluorescent measurement of blood potassium with polymeric optical sensors based on upconverting nanomaterials, *Analytical Chemistry*, 85, 2013, pp. 2617-2622.

[29]. Y. -W. Zhang, X. Sun, R. Si, L. -P. You, C. -H. Yan, Single-crystalline and monodisperse LaF$_3$ triangular nanoplates from a single-source precursor, *Journal of the American Chemical Society*, 127, 2005, pp. 3260-3261.

[30]. Z. Li, Y. Zhang, An efficient and user-friendly method for the synthesis of hexagonal-phase NaYF$_4$: Yb, Er/Tm nanocrystals with controllable shape and upconversion fluorescence, *Nanotechnology*, 19, 2008, 345606.

[31]. X. Wang, J. Zhuang, Q. Peng, Y. Li, A general strategy for nanocrystal synthesis, *Nature*, 437, 2005, pp. 121-124.

[32]. F. Wang, D. K. Chatterjee, Z. Li, Y. Zhang, X. Fan, M. Wang, Synthesis of polyethylenimine/NaYF$_4$ nanoparticles with upconversion fluorescence, *Nanotechnology*, 17, 2006, p. 5786.

[33]. G. Yi, H. Lu, S. Zhao, Y. Ge, W. Yang, D. Chen, L. -H. Guo, Synthesis, characterization, and biological application of size-controlled nanocrystalline NaYF$_4$: Yb, Er infrared-to-visible up-conversion phosphors, *Nano Letters,* 4, 2004, pp. 2191-2196.

[34]. Z. Li, Y. Zhang, Monodisperse silica-coated polyvinylpyrrolidone/NaYF$_4$ nanocrystals with multicolor upconversion fluorescence emission, *Angewandte Chemie*, 118, 2006, pp. 7896-7899.

[35]. T. Zhang, J. Ge, Y. Hu, Y. Yin, A general approach for transferring hydrophobic nanocrystals into water, *Nano Letters*, 7, 2007, pp. 3203-3207.

[36]. A. Dong, X. Ye, J. Chen, Y. Kang, T. Gordon, J. M. Kikkawa, C. B. Murray, A generalized ligand-exchange strategy enabling sequential surface functionalization of colloidal nanocrystals, *Journal of the American Chemical Society*, 133, 2010, pp. 998-1006.

[37]. Z. Chen, H. Chen, H. Hu, M. Yu, F. Li, Q. Zhang, Z. Zhou, T. Yi, C. Huang, Versatile synthesis strategy for carboxylic acid-functionalized upconverting nanophosphors as biological labels, *Journal of the American Chemical Society*, 130, 2008, pp. 3023-3029.

[38]. F. Herranz, M. Morales, A. G. Roca, M. Desco, J. Ruiz-Cabello, A new method for the rapid synthesis of water stable superparamagnetic nanoparticles, *Chemistry-A European Journal*, 14, 2008, pp. 9126-9130.

[39]. H. P. Zhou, C. H. Xu, W. Sun, C. H. Yan, Clean and Flexible Modification Strategy for Carboxyl/Aldehyde-Functionalized Upconversion Nanoparticles and Their Optical Applications, *Advanced Functional Materials*, 19, 2009, pp. 3892-3900.

[40]. M. Wang, J. -L. Liu, Y. -X. Zhang, W. Hou, X. -L. Wu, S. -K. Xu, Two-phase solvothermal synthesis of rare-earth doped NaYF$_4$ upconversion fluorescent nanocrystals, *Materials Letters*, 63, 2009, pp. 325-327.

[41]. N. Bogdan, F. Vetrone, G. A. Ozin, J. A. Capobianco, Synthesis of ligand-free colloidally stable water dispersible brightly luminescent lanthanide-doped upconverting nanoparticles, *Nano Letters*, 11, 2011, pp. 835-840.

[42]. G. Jiang, J. Pichaandi, N. J. Johnson, R. D. Burke, F. C. van Veggel, An effective polymer cross-linking strategy to obtain stable dispersions of upconverting NaYF$_4$ nanoparticles in buffers and biological growth media for biolabeling applications, *Langmuir*, 28, 2012, pp. 3239-3247.

[43]. L. Cheng, K. Yang, S. Zhang, M. Shao, S. Lee, Z. Liu, Highly-sensitive multiplexed in vivo imaging using PEGylated upconversion nanoparticles, *Nano Research*, 3, 2010, pp. 722-732.

[44]. L. Wang, R. Yan, Z. Huo, L. Wang, J. Zeng, J. Bao, X. Wang, Q. Peng, Y. Li, Fluorescence resonant energy transfer biosensor based on upconversion-luminescent nanoparticles, *Angewandte Chemie International Edition*, 44, 2005, pp. 6054-6057.

[45]. C. Wang, L. Cheng, Y. Liu, X. Wang, X. Ma, Z. Deng, Y. Li, Z. Liu, Imaging-Guided Ph-Sensitive Photodynamic Therapy Using Charge Reversible Upconversion Nanoparticles under Near-Infrared Light, *Advanced Functional Materials*, 23, 2013, pp. 3077-3086.

[46]. J. -N. Liu, W. -B. Bu, J. -L. Shi, Silica coated upconversion nanoparticles: A versatile platform for the development of efficient theranostics, *Accounts of Chemical Research*, 48, 2015, pp. 1797-1805.

[47]. A. Sedlmeier, H. H. Gorris, Surface modification and characterization of photon-upconverting nanoparticles for bioanalytical applications, *Chemical Society Reviews*, 44, 2015, pp. 1526-1560.

[48]. M. Wang, G. Abbineni, A. Clevenger, C. Mao, S. Xu, Upconversion nanoparticles: synthesis, surface modification and biological applications, Nanomedicine: Nanotechnology, *Biology and Medicine*, 7, 2011, pp. 710-729.

[49]. J. Zhou, Z. Liu, F. Li, Upconversion nanophosphors for small-animal imaging, *Chemical Society Reviews*, 41, 2012, pp. 1323-1349.

[50]. J. Zhang, B. Li, L. Zhang, H. Jiang, An optical sensor for Cu (II) detection with upconverting luminescent nanoparticles as an excitation source, *Chemical Communications*, 48, 2012, pp. 4860-4862.

[51]. L. Yao, J. Zhou, J. Liu, W. Feng, F. Li, Iridium-Complex-Modified Upconversion Nanophosphors for Effective LRET Detection of Cyanide Anions in Pure Water, *Advanced Functional Materials*, 22, 2012, pp. 2667-2672.

[52]. Y. Ding, H. Zhu, X. Zhang, J. -J. Zhu, C. Burda, Rhodamine B derivative-functionalized upconversion nanoparticles for FRET-based Fe^{3+}-sensing, *Chemical Communications*, 49, 2013, pp. 7797-7799.

[53]. L. Yan, Y. -N. Chang, W. Yin, X. Liu, D. Xiao, G. Xing, L. Zhao, Z. Gu, Y. Zhao, Biocompatible and flexible graphene oxide/upconversion nanoparticle hybrid film for optical pH sensing, *Physical Chemistry Chemical Physics,* 16, 2014, pp. 1576-1582.

[54]. J. Peng, W. Xu, C. L. Teoh, S. Han, B. Kim, A. Samanta, J. C. Er, L. Wang, L. Yuan, X. Liu, High-efficiency in vitro and in vivo detection of Zn2+ by dye-assembled upconversion nanoparticles, *Journal of the American Chemical Society*, 137, 2015, pp. 2336-2342.

[55]. Q. Liu, Y. Sun, T. Yang, W. Feng, C. Li, F. Li, Sub-10 nm Hexagonal Lanthanide-Doped $NaLuF_4$ Upconversion Nanocrystals for Sensitive Bioimaging in Vivo, *Journal of the American Chemical Society*, 133, 2011, pp. 17122-17125.

[56]. W. Zou, C. Visser, J. A. Maduro, M. S. Pshenichnikov, J. C. Hummelen, Broadband dye-sensitized upconversion of near-infrared light, *Nature Photonics,* 6, 2012, pp. 560-564.

[57]. J. Lu, Y. Chen, D. Liu, W. Ren, Y. Lu, Y. Shi, J. Piper, I. Paulsen, D. Jin, One-step protein conjugation to upconversion nanoparticles, *Analytical Chemistry*, 87, 2015, pp. 10406-10413.

[58]. A. E. Guller, A. N. Generalova, E. V. Petersen, A. V. Nechaev, I. A. Trusova, N. N. Landyshev, A. Nadort, E. A. Grebenik, S. M. Deyev, A. B. Shekhter, Cytotoxicity and non-specific cellular uptake of bare and surface-modified upconversion nanoparticles in human skin cells, *Nano Research*, 8, 2015, pp. 1546-1562.

[59]. Z. Li, S. Lv, Y. Wang, S. Chen, Z. Liu, Construction of LRET-based nanoprobe using upconversion nanoparticles with confined emitters and bared surface as luminophore, *Journal of the American Chemical Society*, 137, 2015, pp. 3421-3427.

[60]. M. Kumar, P. Zhang, Highly sensitive and selective label-free optical detection of mercuric ions using photon upconverting nanoparticles, *Biosensors and Bioelectronics*, 25, 2010, pp. 2431-2435.

[61]. M. Kumar, P. Zhang, Highly sensitive and selective label-free optical detection of DNA hybridization based on photon upconverting nanoparticles, *Langmuir*, 25, 2009, pp. 6024-6027.

[62]. P. Alonso-Cristobal, P. Vilela, A. El-Sagheer, E. Lopez-Cabarcos, T. Brown, O. Muskens, J. Rubio-Retama, A. Kanaras, Highly sensitive

DNA sensor based on upconversion nanoparticles and graphene oxide, *ACS Applied Materials & Interfaces*, 7, 2015, pp. 12422-12429.

[63]. Y. Wang, L. Bao, Z. Liu, D. -W. Pang, Aptamer biosensor based on fluorescence resonance energy transfer from upconverting phosphors to carbon nanoparticles for thrombin detection in human plasma, *Analytical Chemistry*, 83, 2011, pp. 8130-8137.

[64]. L. Liu, B. Li, R. Qin, H. Zhao, X. Ren, Z. Su, Synthesis and characterization of new bifunctional nanocomposites possessing upconversion and oxygen-sensing properties, *Nanotechnology*, 21, 2010, 285701.

[65]. S. Liu, L. Zhang, T. Yang, H. Yang, K. Y. Zhang, X. Zhao, W. Lv, Q. Yu, X. Zhang, Q. Zhao, Development of upconversion luminescent probe for ratiometric sensing and bioimaging of hydrogen sulfide, *ACS Applied Materials & Interfaces*, 6, 2014, pp. 11013-11017.

[66]. C. Li, J. Liu, S. Alonso, F. Li, Y. Zhang, Upconversion nanoparticles for sensitive and in-depth detection of Cu^{2+} ions, *Nanoscale*, 4, 2012, pp. 6065-6071.

[67]. R. Han, H. Yi, J. Shi, Z. Liu, H. Wang, Y. Hou, Y. Wang, pH-Responsive drug release and NIR-triggered singlet oxygen generation based on a multifunctional core–shell–shell structure, *Physical Chemistry Chemical Physics*, 18, 2016, pp. 25497-25503.

[68]. J. Liu, Y. Liu, W. Bu, J. Bu, Y. Sun, J. Du, J. Shi, Ultrasensitive nanosensors based on upconversion nanoparticles for selective hypoxia imaging in vivo upon near-infrared excitation, *Journal of the American Chemical Society*, 136, 2014, pp. 9701-9709.

[69]. L. Zhao, J. Peng, M. Chen, Y. Liu, L. Yao, W. Feng, F. Li, Yolk-shell upconversion nanocomposites for LRET sensing of cysteine/homocysteine, *ACS Applied Materials & Interfaces*, 6, 2014, pp. 11190-11197.

[70]. H. Li, L. Wang, $NaYF_4$: Yb^{3+}/Er^{3+} nanoparticle-based upconversion luminescence resonance energy transfer sensor for mercury (II) quantification, *Analyst*, 138, 2013, pp. 1589-1595.

[71]. J. Liu, Y. Liu, Q. Liu, C. Li, L. Sun, F. Li, Iridium (III) complex-coated nanosystem for ratiometric upconversion luminescence bioimaging of cyanide anions, *Journal of the American Chemical Society*, 133, 2011, pp. 15276-15279.

[72]. P. Zhang, S. Rogelj, K. Nguyen, D. Wheeler, Design of a highly sensitive and specific nucleotide sensor based on photon upconverting particles, *Journal of the American Chemical Society*, 128, 2006, pp. 12410-12411.

[73]. F. Zhou, U. J. Krull, Spectrally matched duplexed nucleic acid bioassay using two-colors from a single form of upconversion nanoparticle, *Analytical Chemistry*, 86, 2014, pp. 10932-10939.

[74]. M. Kumar, Y. Guo, P. Zhang, Highly sensitive and selective oligonucleotide sensor for sickle cell disease gene using photon upconverting nanoparticles, *Biosensors and Bioelectronics*, 24, 2009, pp. 1522-1526.

[75]. F. Yuan, H. Chen, J. Xu, Y. Zhang, Y. Wu, L. Wang, Aptamer-Based Luminescence Energy Transfer from Near-Infrared-to-Near-Infrared Upconverting Nanoparticles to Gold Nanorods and Its Application for the Detection of Thrombin, *Chemistry-A European Journal*, 20, 2014, pp. 2888-2894.

[76]. H. Chen, F. Yuan, S. Wang, J. Xu, Y. Zhang, L. Wang, Aptamer-based sensing for thrombin in red region via fluorescence resonant energy transfer between $NaYF_4$: Yb, Er upconversion nanoparticles and gold nanorods, *Biosensors and Bioelectronics*, 48, 2013, pp. 19-25.

[77]. S. Xu, B. Dong, D. Zhou, Z. Yin, S. Cui, W. Xu, B. Chen, H. Song, Paper-based upconversion fluorescence resonance energy transfer biosensor for sensitive detection of multiple cancer biomarkers, *Scientific Reports*, 6, 2016, p. 23406.

[78]. E.-J. Jo, H. Mun, M.-G. Kim, Homogeneous immunosensor based on luminescence resonance energy transfer for glycated hemoglobin detection using upconversion nanoparticles, *Analytical Chemistry*, 88, 2016, pp. 2742-2746.

[79]. C. Liu, Z. Wang, H. Jia, Z. Li, Efficient fluorescence resonance energy transfer between upconversion nanophosphors and graphene oxide: a highly sensitive biosensing platform, *Chemical Communications*, 47, 2011, pp. 4661-4663.

[80]. C. Zhang, Y. Yuan, S. Zhang, Y. Wang, Z. Liu, Biosensing platform based on fluorescence resonance energy transfer from upconverting nanocrystals to graphene oxide, *Angewandte Chemie International Edition*, 50, 2011, pp. 6851-6854.

[81]. R. Deng, X. Xie, M. Vendrell, Y.-T. Chang, X. Liu, Intracellular glutathione detection using MnO_2-nanosheet-modified upconversion nanoparticles, *Journal of the American Chemical Society*, 133, 2011, pp. 20168-20171.

[82]. Y. Wang, P. Shen, C. Li, Y. Wang, Z. Liu, Upconversion fluorescence resonance energy transfer based biosensor for ultrasensitive detection of matrix metalloproteinase-2 in blood, *Analytical Chemistry*, 84, 2012, pp. 1466-1473.

[83]. S. Wu, N. Duan, X. Ma, Y. Xia, H. Wang, Z. Wang, Q. Zhang, Multiplexed fluorescence resonance energy transfer aptasensor between upconversion nanoparticles and graphene oxide for the simultaneous determination of mycotoxins, *Analytical Chemistry*, 84, 2012, pp. 6263-6270.

[84]. S. Jiang, Y. Zhang, Upconversion nanoparticle-based FRET system for study of siRNA in live cells, *Langmuir*, 26, 2010, pp. 6689-6694.

[85]. S. Li, L. Xu, W. Ma, X. Wu, M. Sun, H. Kuang, L. Wang, N. A. Kotov, C. Xu, Dual-mode ultrasensitive quantification of microRNA in living cells by chiroplasmonic nanopyramids self-assembled from gold and upconversion nanoparticles, *Journal of the American Chemical Society*, 138, 2016, pp. 306-312.

[86]. P. Vilela, A. El-Sagheer, T. M. Millar, T. Brown, O. L. Muskens, A. G. Kanaras, Graphene Oxide-Upconversion Nanoparticle Based

Optical Sensors for Targeted Detection of mRNA Biomarkers Present in Alzheimer's Disease and Prostate Cancer, *ACS Sensors*, 2, 2017, pp. 52-56.

[87]. M. Wang, W. Hou, C. -C. Mi, W. -X. Wang, Z. -R. Xu, H. -H. Teng, C.-B. Mao, S. -K. Xu, Immunoassay of goat antihuman immunoglobulin G antibody based on luminescence resonance energy transfer between near-infrared responsive NaYF$_4$: Yb, Er upconversion fluorescent nanoparticles and gold nanoparticles, *Analytical Chemistry*, 81, 2009, pp. 8783-8789.

[88]. H. S. Mader, O. S. Wolfbeis, Optical ammonia sensor based on upconverting luminescent nanoparticles, *Analytical Chemistry*, 82, 2010, pp. 5002-5004.

[89]. H. Chen, H. Li, J. -M. Lin, Determination of ammonia in water based on chemiluminescence resonance energy transfer between peroxymonocarbonate and branched NaYF$_4$: Yb^{3+}/Er^{3+} nanoparticles, *Analytical Chemistry*, 84, 2012, pp. 8871-8879.

Chapter 3

A Review of Nano-Particle Analysis with the PAMONO-Sensor

Jan Eric Lenssen, Victoria Shpacovitch, Dominic Siedhoff, Pascal Libuschewski, Roland Hergenröder and Frank Weichert

3.1. Introduction

Surface plasmon resonance (SPR) is a highly sensitive optical method which allows studying interactions between different types of biomolecules: peptides, proteins, nucleic acids. In this review we focus on the hardware and software features of a sensor that applies this optical method: the PAMONO-sensor (plasmon assisted microscopy of nano-objects) that is shown in Fig. 3.1, together with an Odroid XU3 for mobile data processing.

In the conventional SPR approach, the layers of biomolecules bound to a noble metal - usually gold - sensor surface alter the refractive conditions near the sensor surface. These changes provide sufficient information for real-time measurements of the biomolecule binding efficiency. Thus, the determination of affinity constants or concentration measurements are common tasks which can be solved with an SPR sensor. However, for a long time the detection and quantification of individual nano-particles remained an unsolvable problem for plasmonic biosensors. The limitations were thought to be associated with the problem of the lateral resolution. It was declared that the length of

Jan Eric Lenssen
Department of Computer Science VII, Dortmund University of Technology, Dortmund, Germany

plasmonic propagation is crucial for the size resolution of bound particles. Thus, the minimal size of objects detected by the SPR-based sensor might be limited to a micrometer scale. Recently, however, independent research teams proved the power of plasmonic sensors in the quantification and analysis of biological nanometer scale particles [1-5].

Fig. 3.1. Prototype of the PAMONO-sensor together with an Odroid XU-3 in the bottom right corner, on which the data analysis is performed.

In Section 3.2, we give a short overview of state-of-the-art technical aspects of the PAMONO-sensor. Subsequently, we discuss the ability to detect signals of individual viral or nonbiological particles and to determine the concentration of particles in analyzed samples. In Section 3.3 we focus on the aspects that are highlighted within the analysis-architecture overview, cf. Fig. 3.3. This consists of the sensor data acquisition, feature extraction, classification and optimization of the whole pipeline to match specific characteristics of the PAMONO-sensor and the sensor data. In Section 3.4, we outline selected results of the different approaches. Finally, we summarize the results and describe future work in Section 3.5.

3.2. Technical Aspects of the PAMONO-Sensor

A demonstrated prime advantage of the PAMONO-sensor [1, 8, 17] is its ability to detect signals attributed to individual viral particles or non-biological particles of different sizes down to 40 nm in diameter. Thus, this sensor provides instant information about the existence of intact biological nano-particles or nano-vesicles in the analyzed sample. The first images of individual nano-particle binding events to the PAMONO-sensor surface are presented in the work of Zybin et al. [1].

The PAMONO-sensor performs imaging of bound sub-micrometer and nano-particles using a Kretschmann's scheme [6] of plasmon excitation. The feasibility of this detection is explained in the recent theoretical model described by Zybin et al. [7]. Fig. 3.2 shows the functionality of the PAMONO-senor. By illumination of a gold sensor surface through a glass prism near the resonance, plasmon waves are excited in the gold layer. The PAMONO-sensor deals with the detection of particles, which are much smaller than the plasmon wavelength and cause only a small disturbance of the primary plasmon wave. Thus, the front of primary excited plasmon waves remains practically undisturbed. Nevertheless, a subset of primary plasmon waves can be scattered and can form local and weak, secondary concentric plasmonic waves around the bound nano-particles. These secondary plasmonic waves can be disclosed by a Charge-Coupled Device (CCD) detector and therefore help to visualize the binding of individual nano-vesicles [7].

Further research described analytical features of the PAMONO-sensor [3, 8]. Gurevich et al. investigated the ability of the PAMONO-sensor to measure particle concentrations in aqueous samples [3]. They used different concentrations of 200 nm polystyrene particles and reveal a linear dependency between the number of signals detected by the PAMONO-sensor and the concentration of 200 nm polystyrene particles in the diluted samples. This calibration curve and theoretical assumptions helped to estimate the concentration of Virus-Like Particles (VLPs) in the analyzed sample. The prediction of the VLP concentration utilizing PAMONO-sensor measurements was in good agreement with the concentration measurements with the Enzyme-Linked Immunosorbent Assay (ELISA) method for the same samples. This result opens an outlook for the determination of particle concentration without preliminary calibration of the PAMONO-sensor. Noteworthy, in case of medium concentration values, the determination of particle concentration by the PAMONO-sensor requires only minutes. In contrast, the ELISA method takes hours to provide information about VLP concentration in a sample, independent of the concentration range.

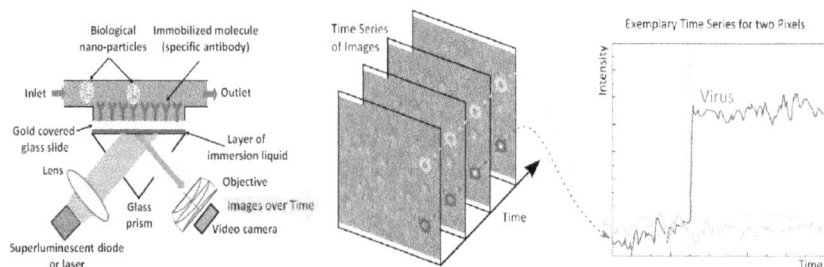

Fig. 3.2. This scheme describes a set-up of the PAMONO-sensor, which can be used for the detection and quantification of biological nano-particles. A gold covered glass slide serves as a highly sensitive sensor surface, which can be functionalized with target specific antibody or other molecules, e.g. mucins, that ensure the desired selectivity. Immobilization of the targetspecific antibody on the sensor surface is achieved via formation of self-assembling monolayers (SAMs) such as thiolated biotin-streptavidin-antibody conjugated with biotin. Other SAMs can be also used. The CCD detector produces image sequences in which temporal intensity steps can be observed at positions of immobilized particles. Adapted from [10].

In the same work Gurevich et al. discussed the lower concentration detection limit of the PAMONO-sensor and ways to improve it [3]. According to the calculations performed by the authors, the low concentration of 8×104 particles per ml can be determined by the PAMONO-sensor during 1 hour measurement time and on 0.01 cm^2 of the sensor surface area. The detection limit can be improved by enlarging the sensor detection area and by increasing the analysis time. For example, a sensor surface area can be enlarged up to 0.1 cm^2, reducing the imaging magnification via an application of a camera with smaller pixel size and higher resolution [3]. The work of Shpacovitch et al. [8] validates the bioanalytical characteristics of the PAMONO-sensor. In this study Tobacco-Mosaic Virus (TMV, 15 nm –18 nm diameter and 300 nm height), inactivated Influenza A Virus (IAV, spherical with 80 nm –120 nm in diameter) as well as Human Immunodeficiency Virus VLPs (HIV-VLPs, spherical with 100 nm –140 nm in diameter) were used. This study demonstrated that the shape of investigated nano-vesicles may be important. Indeed, the PAMONO-sensor enabled the detection of IAV and HIVVLPs binding, but failed to visualize the binding of TMV [8]. In addition, the results of this study revealed high specificity of the detected signals. Moreover, this specificity is primarily connected with the specificity of anti-target antibody used for the sensor functionalization. The authors also show that the PAMONO-sensor can detect individual viral particles in solutions containing serum.

Under certain circumstances the information about quantity and characteristics of intact biological vesicles serves as a basis for disease diagnostic. In this case, just knowing the particle content (proteins or genetic material) is not sufficient since intact particles serve as a means of inter-cellular communication. An example for this are extracellular vesicles (EVs): exosomes and microvesicles (MVs). It was checked whether the PAMONO-sensor enables the detection and quantification of MVs derived from eukaryotic cells. The study addressing this question was recently performed by Shpacovitch et al. [16]. The authors demonstrated the applicability of the PAMONO-sensor for the detection of MVs in aquatic samples. Moreover, the sensor helped to quickly compare relative concentrations of microvesicles between samples [16]. In addition, novel machine learning techniques were employed in order to perform particle size distribution analysis [16], which is further discussed in Section 3.3.2. Taken together, these findings can help to create a sensor platform for quantification and size distribution profiling of microvesicles. In turn, this information is very valuable for the estimation of physiological or pathological changes in eukaryotic organisms.

Another challenging task is to achieve a real-time analysis of the sensor data with a high detection rate and a low false detection rate. This includes an automatic classification and separation of viruses from artifacts. In the following Section 3.3 an automatic image processing approach is presented, which provides high detection quality, real-time capability, and energy efficiency suitable for a mobile biosensor.

3.3. Processing of PAMONO Sensor Data

The PAMONO-sensor produces image sequences with low signal-to-noise ratio (low-SNR) which can be analyzed by methods from digital image processing. The following chapter outlines methodical aspects which are important for extracting information like existence, number and size of particles. First, the General-Purpose Computing on Graphics Processing Units (GPGPU) real-time image processing pipeline is presented in Section 3.3.1. Techniques for particle size classification are outlined in Section 3.3.2, software parameter optimization techniques in Section 3.3.3, while hardware optimization and analysis is surveyed in Section 3.3.4. Selected results are presented in Section 3.4, subsequently.

3.3.1. Detection, Classification and Quantification of Nano-Objects

The analysis of PAMONO-sensor data is done by a real-time capable image processing pipeline that uses GPGPU. An overview of the whole pipeline is presented in Fig. 3.3. The pipeline can be separated into four parts: preprocessing, pattern detection, feature extraction, and classification. First, the sensor image sequence data, for which an example image is shown in Fig. 3.4 (a), is preprocessed through removing the dominating constant background, followed by intensity normalization and noise reduction algorithms [10]. For background elimination, the current image is divided by the mean image of an image set. Whether the set is constant for every image or filled dynamically with a number of previous frames is an adjustable parameter. The division is a consequence of the physically motivated signal model that describes an image sequence $I(x, y, t)$ as:

$$I(x,y,t) = B(x,y) \cdot A(x,y,t) \cdot P(x,y,t) + N(x,y,t), \quad (1.1)$$

which was proposed by Siedhoff et al. [18]. In this equation, $B(x,y)$ is the constant background signal, $A(x,y,t)$ an unknown artifact signal, $P(x,y,t)$ the target particle signal and $N(x,y,t)$ a random noise term. Thus, dividing $I(x,y,t)$ through an approximated $B(x,y)$ leads to images containing the target signal, artifacts and noise. $B(x,y)$ can be approximated by computing the average of an image subsequence without particle signals. Noise reduction is performed before and after background elimination and intensity normalization. Before, temporal noise is reduced by taking an element-wise median of an image subsequence. After the normalization, the spatial and temporal noise is further reduced by a combination of standard image filters and wavelet denoising [11]. A preprocessed image is shown in Fig. 3.4 (b).

The preprocessed signal is further processed by the pattern detector step of the pipeline that detects nano-particle candidates by time-series pattern matching [10], creating local object hypotheses. An improvement of the pattern detector with fuzzy-rules for pixel classification has been presented and evaluated by Libuschewski et al. [11, 14]. Siedhoff et al.

86

have presented an alternative approach for detecting particles using translation-invariant wavelet features [19], which is not yet real-time capable. As an additional alternative, a fully convolutional neural network for pixel classification was applied as a pattern detector [15], which is also real-time capable. A comparison of the different approaches is shown in Section 3.4.1.

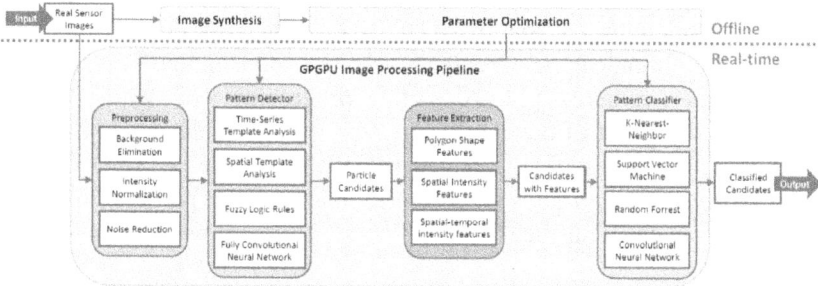

Fig. 3.3. The GPGPU processing pipeline consisting of four main steps: preprocessing of images, pattern detection, candidate feature extraction, and classification. Algorithm parameters can be set by an offline parameter optimization step that uses synthetic sensor data for model training. Adapted from [20].

(a) Nearly constant raw sensor signal

(b) Pre-processed image with eliminated constant background

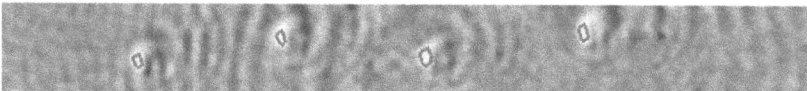

(c) Result of the detection with identified particles

Fig. 3.4. PAMONO sensor images from three different steps of the pipeline: (a) shows the raw image before processing; (b) the pre-processed image after background elimination, and (c) detected particles that were found in the image. Contrast of the images is enhanced for visualization purposes.

The result of this step is a binary mask that marks each pixel that contains a part of the target particle and maybe artifact signal. Regions of marked pixels are processed with morphological operations before generating two-dimensional candidate polygons using the marching square algorithm [23]. After candidate generation, the candidate polygons (cf. Fig. 3.4 (c)) are classified as correct or false detections. This step sorts out false-positives, i.e. detected artifact parts or noise peaks. Extracted features for classification are based on polygon shape as well as spatial and temporal intensity features [10]. Different classifiers, namely k-Nearest-Neighbor, Support Vector Machines, and Random Forests were trained and evaluated [24-26]. After processing one whole image sequence, the result of the classifying step is a set of detected particles, from which the classifier removed (ideally) all false-positives but no true-positives. This allows the quantification of the particles that are bound to the sensor surface.

3.3.2. Particle Size Distributions

The size of a bound particle has direct impact on the resulting signal intensity in the captured images. Furthermore, the dependency between particle size and signal intensity is nearly linear [1]. Therefore, the particle size can be inferred by applying image processing methods, if a constant refractive index is given. Lenssen et al. performed Experiments using a convolutional neural network for probabilistic classification, which receives small image patches containing particle candidates as input and produces confidences for different size intervals [15]. The most probable particle size can be inferred by computing the expectancy value over all confidence values. For a whole image sequence, a particle size distribution can be estimated by counting the particles of each size [16]. The neural network was trained using synthetic image patches which were generated by averaging over a randomly sampled subset of real image patches, where each patch contains one particles [15]. The set of patches from which the subset is sampled contains images of particles for which the estimated mean size but not the exact size is known. Such sets can be obtained by perform PAMONO-sensor measurements of liquid samples for which the mean size of all contained particles is known. Furthermore, training images for some sizes have been interpolated from two different sizes so measurements only have to be performed for some particle sizes. In addition, the synthetic training images were augmented by strong artificial Gaussian noise.

3.3.3. Software Parameter Optimization

Several parts of the pipeline, namely the preprocessing, pattern detector and classifier can be adjusted by different parameters. Physical properties of the sensor's surface, the reflected light, and external conditions, like the sensor's temperature, influence the resulting sensor data, leading to different parameter sets required for optimal particle detection results. An offline optimization routine that uses synthetic sensor data for finding optimal parameter sets for every setup, as implied by Fig. 3.3, was developed by Siedhoff [18]. First, synthetic data sets are generated using the proposed signal model and a small amount of manually labeled real data. Real background and artifact signals are randomly combined with a small amount but augmented, manually labeled target signals and random noise, providing nearly arbitrary amounts of different data sets with automatically created ground truth. Then, these synthetic data sets are utilized as training data for a genetic algorithm that approximates an optimal parameter set for the given data [18].

3.3.4. Miniaturization and Resource-Aware Optimizations

Miniaturization of the sensor to design a mobile nano-object detector also comes with challenges for software and hardware design. Mobile graphic processing units (GPUs), which are needed to apply the presented real-time nano-object detection pipeline, require a lot of energy when computing highly parallel image processing tasks. In order to evaluate different GPU architectures, an energy-aware design space exploration has been done by Libuschewski et al. [12]. Different classification and general image processing algorithms were simulated for different GPUs, to explore the energy consumptions of methods for particle detection. As an alternative to reduce power consumption in mobile sensors, offloading can be used. Some of the computing tasks are sent to a resourceful server that can calculate results without energy constraints and with higher computing power. A multi-objective computation offloading framework with Long-Term Evolution (LTE) as mobile communication standard has been developed and evaluated by Libuschewski et al. [13]. The framework consists of a multi-objective decision making process that is based on several constraints. The objectives are energy/power consumption, number of core cycles and run time. In order to conclude hardware and software optimization for PAMONO image data, a combined study was performed, optimizing software and hardware parameters with respect to energy consumption, run-time, and detection quality [21] using the NSGA-II genetic algorithm. The goal was to investigate the impact of different parameter variations as well as finding Pareto-optimal software and hardware setups for the proposed pipeline.

3.4. Results and Discussion

This section presents selected PAMONO sensor data analysis results that were achieved with the outlined methodology. We begin with summarizing particle quantification and software parameter optimization in Section 3.4.1. In Section 3.4.2 particle size classification results are given. Then, we present results of different hardware and offloading optimizations as well as results from the combined study in Section 3.4.3.

3.4.1. Measurements of Detection Quality using Optimized Pipeline Parameters

For evaluation of the real-time pipeline as well as substantiating the proposed signal model and image synthesis procedure, Siedhoff presented results for a PAMONO-sensor data set containing 200 nm polystyrene particles [18]. The 4000 images have a resolution of 1080 px ×145 px and were captured with the 12 bit CCD camera *Prosilica GC 2450*. Recall and precision values for synthetic training, synthetic test, and real test data have been evaluated. All used synthetic and real data sets are available online for reproduction. From a small subset of the real data set, disjunct synthetic training and test data sets were synthesized. The synthetic training data set was used to optimize parameters in respect to the recall value. Then, the optimal parametrized pipeline is evaluated with the synthetic training, test, and with the manually labeled, unused part of the real data set. The results are given in Fig. 3.5 (a) [18]. It can be observed that the parametrized pipeline achieves similar results on the real data set, validating the image synthesis procedure.

The alternative method using translation invariant wavelets (c.f. Section 3.3.1) was compared to the evaluated approach with fuzzy template matching by Siedhoff et al. [20]. The experiments were performed for synthetic pixel time-series that were extracted from three different datasets containing different particles: 200 nm particles in high quality (higher SNR), 200 nm particles in low quality (lower SNR), and 100 nm particles in high quality. Here, class balance is ensured so accuracy is used as quality measure. The mean results are compared in Fig. 3.5 (b). Besides the better detection performance, a better robustness regarding noise could be observed. Adding more synthetic noise to the data affects accuracy only by a small margin. Multiplying the random Gaussian noise applied to the data with five leads to a drop in accuracy of less than one percent [20].

The optimized fuzzy template matching was also compared to the fully convolutional network from [15]. The results are shown in Fig. 3.5 (c). It can be observed, that the optimized fuzzy template matching produces slightly better results in terms of recall and precision. It has to be noted though that the parameter optimization for fuzzy template matching has to be performed for each data set individually before these results can be achieved [20]. The fully convolutional network, in contrast, is always the same for each data set. It was neither retrained nor was its architecture revised over the examined data sets.

3.4.2. Particle Size Classification

We present results of the particle size classification approach using a convolutional neural network as presented by Lenssen et al. [15], as well as resulting particle size distributions [16]. First, the accuracy of the classification is presented. The number of classes is 23, which represent 10 nm intervals from 80 nm to 300 nm. Since the evaluation was performed using real test image patches, which were captured by analyzing a liquid sample with known mean particle size with the PAMONO-sensor, there is no exact ground truth for each example. For a set of image patches, only the estimated mean particle size is given by the sample manufacturer. The ground truths are chosen as these means and therefore are noisy. To take this, as well as sensor and imaging inaccuracies, into account, a soft accuracy

$$a^r = \frac{\left|\{e \in \mathcal{E} \mid |c^{pred}(e) - c^{corr}(e)| \le r\}\right|}{|\mathcal{E}|}, \tag{1.2}$$

is evaluated. It counts the predictions which are in the interval of size $2 \cdot r$ around the correct class and therefore enables to evaluate, whether the predicted sizes are near the ground truths or far away.

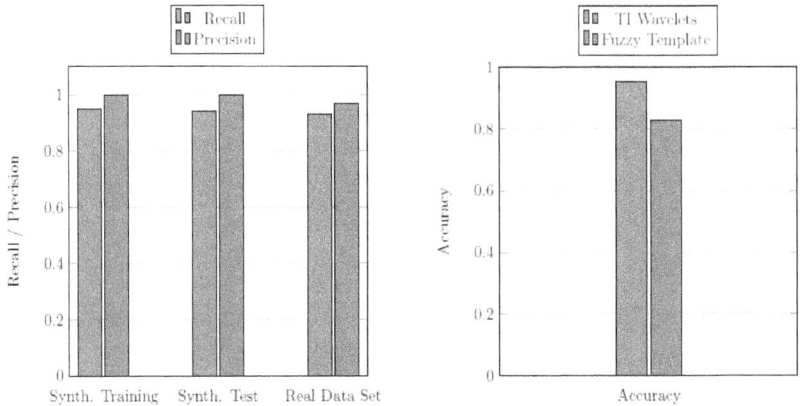

(a) Synthetic vs. real data

(b) Wavelets vs. fuzzy templates

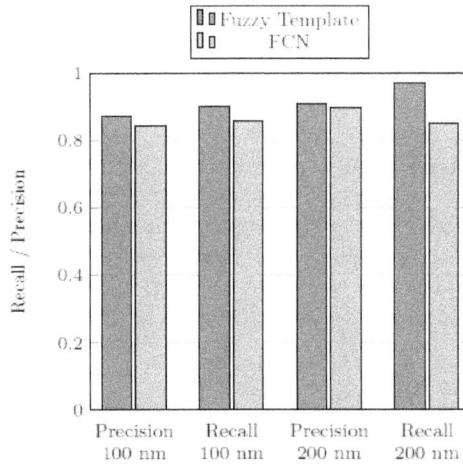

(c) Fuzzy templates vs. FCN

Fig. 3.5. Selected experiments from different research work: (a) shows detection results of the analysis pipeline applied on PAMONO-sensor data with 200 nm polystyrene particles of an experiment performed in [18] and compares the results for synthetic and real sensor data, (b) shows the fuzzy-improved template matching versus the not real-time capable translation invariant wavelet detection from [20] while (c) compares the fuzzy-improved template matching with the fully convolutional network from [15] using two different data sets (containing 100 nm and 200 nm particles, respectively). The three experiments a), (b) and (c) were performed under different circumstances and on different data sets. Figs. (a) and (b) adapted from [20].

Fig. 3.6 (a) shows the a^r for synthetic and real test data sets and different r, each set containing 2500 examples per class.

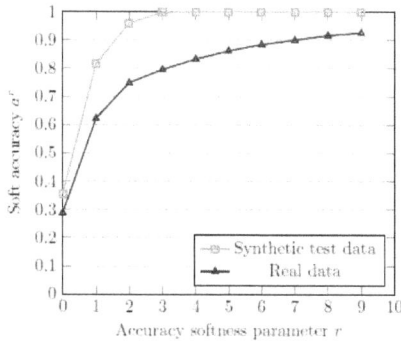

(a) Soft accuracy

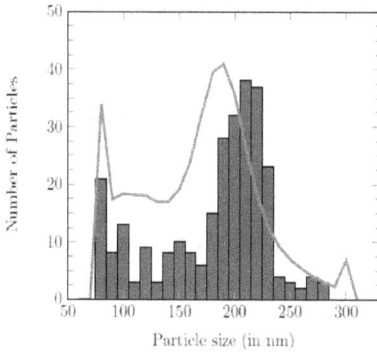

(b) 80 nm and 200 nm data set

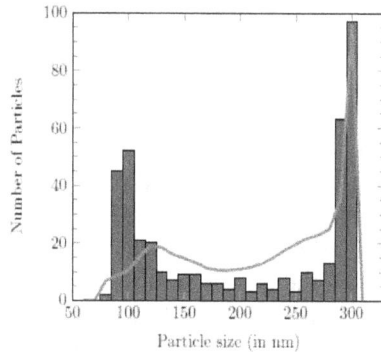

(c) 100 nm and 300 nm data set

Fig. 3.6. Results for particle size classification and size distributions: (a) shows the soft accuracy values for the particle size classification for increasing r [15] while (b) and (c) show resulting particle size distributions for two different data sets. According to the reference measurements, the first data set contains mainly 80 nm and 200 nm particles while the second one contains 100 nm and 300 nm particles. The reference measurements performed with a Malvern LM10 device are shown in red. Fig. (a) adapted from [15].

In addition, examples of estimated particle size distributions resulting from new experiments are shown in Fig. 3.6 (b) and 1.6 (c) and are compared to reference measurements performed by a Malvern LM10 device that were binned into the interval between 80 nm and 300 nm and normalized. The authors observed errors in accuracy, indicated by a bias in the size dimension of the distribution, but noted that the network shows higher precision than the reference measurements [16]. Overall, meaningful information, which leads to plausible distributions, could be automatically extracted.

3.4.3. Energy-Aware Optimization and Offloading

As described in Section 3.3.4, the miniaturization of the PAMONO-sensor is an ongoing goal of the sensor related research. The data processing algorithms shall be executed on small, battery-powered mobile devices and should make use of the possibility to delegate tasks to more powerful computing devices around them, which is called offloading. Therefore, efficiency in energy, memory and computing power consumption is important.

In the following we present selected results from the different studies that were made with respect to offloading and combined hardware/software optimization. First, we outline the possible energy savings of offloading computation from mobile image processing devices [12, 13]. Four different mobile GPUs were simulated using in four different offloading schedules: NVIDIA Geforce GTX 520M, 540M, 560M, and 580M for the schedules *no offloading* (1), *offloading after pattern detector* (2), *offloading after preprocessing* (3), and *full offloading* (4). The schedule names describe the point in the pipeline at which the data is sent to a remote server that continues the computation and returns the result. Evaluated energy consumptions and run times for parameter optimized pipeline executions with offloading are shown in Table 3.1 for the different GPUs and schedules. The outlined energy consumption measurements contain the energy consumptions from the GPU as well as from the LTE communication. Some of the GPUs were not able to compute all parts of the pipeline resulting in some missing combinations. The results show a trade-off between energy consumption and run time that can be adjusted. With full offloading the energy consumption can be reduced to 3.51 J while having a high run time, leading to non real-time computation durations. With no offloading, the run time can be reduced to 0.51 s with the downside of having a high energy consumption.

As described in Section 3.3.4, a combined study of hardware and software optimization for the featured PAMONO image analysis pipeline was made by Neugebauer et al. [21]. Three different experiments were performed, one with software parameter optimization only, one with mobile hardware configuration optimization only and one with both at the same time. The optimized mobile hardware parameters consist of energy mode, core frequency, GPU work group sizes, and memory allocation size while software parameters are given by the analysis pipeline's algorithms. As experiment data, two synthetic data sets, one for training and one for testing, with 1000 images each, a resolution of 706 px ×167 pixels and 200 nm particles were used.

Table 3.1. Offloading experiment results obtained by Libuschewski et al. [13]. Each row shows the energy consumption and run time for one pipeline execution under different hardware/software parameters.

Schedule	GPU	Energy consumption	Run time
1: no offloading	560M	35.72 J	0.51 s
	580M	65.18 J	0.56 s
2	520M	58.0 J	2.82 s
	540M	35.54 J	1.49 s
	560M	15.22 J	0.87 s
	580M	23.21 J	0.85 s
3	520M	34.76 J	25.67 s
	540M	21.89 J	24.94 s
	560M	9.47 J	24.59 s
	580M	15.66 J	24.6 s
4: full offloading	None	3.51 J	27.82 s

The data sets are publicly available [22]. Fig. 3.7 shows the Pareto fronts of the multiobjective optimization as well as the dominated sets for the third experiment, the combined optimization of software and hardware. A few selected models from all three experiments are detailed in Table 3.2. Compared to the baseline parameter set that was obtained by optimizing only in respect of detection quality, large gains can be observed. Without quality loss, parameter sets that result in 76 % energy savings and a four times speed-up were found while optimizing with respect to all three objectives. With minor loss in quality, 93 % energy savings and an eight times speed-up is possible. Comparing the different experiments, it can be observed that optimizing software and hardware target parameters simultaneously leads to better results than just optimizing one of them. In addition, the hardware only parameter optimization is able to save energy but lead only to minor improvements in execution time.

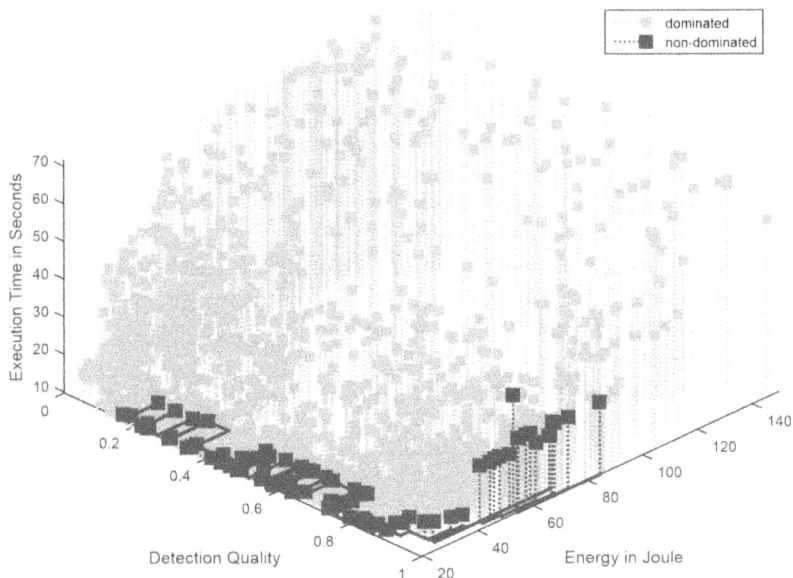

Fig. 3.7. Pareto front and dominated models from the sensor pipeline's software/hardware optimization experiments performed in [21] with objectives of energy consumption, execution time and detection quality. Pareto-optimal individuals are marked blue. [21]

Table 3.2. Selected parameter sets from the multi-objective optimization performed in [21]. The baseline model is the result of using only detection quality as objective. The other rows contain individuals of multi-objective optimizations with different optimization target parameters.

Optimization target	Detection quality	Energy consumption	Run time
Baseline	0.995	370.0 J	119.8 s
Hardware	0.995	233.5 J	118.9 s
	0.995	344.6 J	116.2 s
Software	0.995	87.2 J	29.7 s
	0.883	59.7 J	20.6 s
	0.723	43.9 J	14.7 s
	0.400	34.4 J	12.0 s
Both	0.995	57.5 J	29.3 s
	0.878	27.7 J	14.8 s
	0.649	33.6 J	10.4 s
	0.558	33.0 J	10.0 s

3.5. Conclusion and Outlook

In this chapter, we have inspected a selection of relevant aspects concerning PAMONO-sensor development as well as the real time-capable and energy-aware processing of PAMONO-sensor data in a portable device. After describing recent experiments regarding different particle types, we tackled the aspect of how to detect and quantify nano-particles with the PAMONOsensor automatically. Furthermore, we introduced a generalized processing pipeline for the PAMONO sensor data – sub-steps are preprocessing, pattern detection, feature extraction, and classification. Moreover, we demonstrated how resource-aware optimizations can significantly improve energy consumption, execution time, detection quality, or a trade-off of the aforementioned on a mobile biosensor.

These issues make the PAMONO-sensor a promising candidate for the development of a sensor platform for the real-time characterization and comparison of different EV types as bio-markers of physiological and pathological changes in organisms. Extracellular vesicles (EVs) are shed or released by a majority of eukaryotic and prokaryotic cells. EVs can carry nucleic acids, peptides and other active molecules as a cargo. Acting this way, EVs serve as a means of protected inter-cellular communication and support the interchange of "information" between cells. EVs can be divided in two groups: exosomes (20 nm - 100 nm) and microvesicles (100 nm - 1000 nm). Both groups participate in different functions of cells and organisms and, thus, can reflect physiological and pathological changes in organisms. However, actual and commonly used methods for characterization and quantification of EVs – flow cytometry and nano-particle tracking analysis – have serious limitations and require labeling of EVs for specific detection [9]. Plasmonic sensors proved to overcome these constrains [5]. However, the complexity of the sensor chip preparation used in this study [5] requires long time and limits the size of nano-vesicles, which can be detected simultaneously on the same sensor chip. Further work includes testing the ability of the PAMONO-sensor to quantify and characterize EVs. Indeed, the power of the PAMONO-sensor in quantification of EVs and its ability to provide particle size profiling of the analyzed mixtures of non-organic particles were evidenced [16]. These findings make the PAMONO-sensor a potential candidate for the quantification and analysis of size distribution pattern of EVs in liquid biopsy samples.

Acknowledgment

Part of the work on this chapter has been supported by Deutsche Forschungsgemeinschaft (DFG) within the Collaborative Research Center SFB 876 Providing Information by Resource Constrained Analysis, project B2. The authors are also deeply grateful to Dr. A. Zybin for the proofreading of the manuscript.

References

[1]. Zybin, Y. A. Kuritsyn, E. L. Gurevich, V. V. Temchura, K. Überla, K. Niemax, Realtime detection of single immobilized nano-particles by surface plasmon resonance imaging, *Plasmonics*, 5, 2010, pp. 31–35.

[2]. S. P. Wang, X. N. Shan, U. Patel, X. P. Huang, J. Lu, J. H. Li, N. J. Tao, Label-free imaging, detection, and mass measurement of single viruses by surface plasmon resonance, *Proc. Natl. Acad. Sci. USA*, 107, 2010, pp. 16028–16032.

[3]. E. L. Gurevich, V. V. Temchura, K. Überla, A. Zybin, Analytical features of particle counting sensor based on plasmon assisted microscopy of nano-objects, *Sensors and Actuators B: Chemical,* 160, 2011, pp. 1210–1215.

[4]. A. R. Halpern, J. B. Wood, Y. Wang, R. M. Corn, Single-nanoparticle near-infrared surface plasmon resonance microscopy for real-time measurements of DNA hybridization adsorption, *ACS Nano*, 8, 2014, pp. 1022–1030.

[5]. H. Im, H. Shao, Y. I. Park, V. M. Peterson, C. M. Castro, R. Weissleder, H. Lee, Labelfree detection and molecular profiling of exosomes with a nano-plasmonic sensor, *Nature Biotechnology*, 32, 5, 2014, pp. 490–495.

[6]. E. Kretschmann, Determination of optical constants of metals by excitation of surface plasmons, *Z. Physics*, 241, 1971, pp. 313–324.

[7]. Zybin, V. Shpacovitch, J. Skolnik, R. Hergenröder, Optimal conditions for SPR-imaging of nano-objects, *Sensors and Actuators B: Chemical,* 239, 2017, pp. 338–342.

[8]. V. Shpacovitch, V. Temchura, M. Matrosovich, J. Hamacher, J. Skolnik, P. Libuschewski, D. Siedhoff, F. Weichert, P. Marwedel, H. Müller, K. Überla, R. Hergenröder, A. Zybin, Application of surface plasmon resonance imaging technique for the detection of single spherical biological sub-micrometer particles, *Analytical Biochemistry,* 486, 2015, pp. 62– 69.

[9]. U. Erdbrugger, J. Lannigan, Analytical challenges of extracellular vesicle detection: A comparison of different techniques, *Cytometry. Part A: The Journal of the International Society for Analytical Cytology*, 89, 2, 2016, pp. 123–134.

[10]. D. Siedhoff, F. Weichert, P. Libuschewski, C. Timm, Detection and Classification of Nano Objects in Biosensor Data, in *Proceedings of the*

Microscopic Image Analysis with Applications in Biology Conference, 2011, pp. 1-6.

[11]. P. Libuschewski, D. Siedhoff, C. Timm, A. Gelenberg, F. Weichert. Fuzzy-enhanced, Real-time capable Detection of Biological Viruses Using a Portable Biosensor, in *Proceedings of the International Joint Conference on Biomedical Engineering Systems and Technologies (BIOSIGNALS)*, 2013, pp. 169-174.

[12]. P. Libuschewski, D. Siedhoff, F. Weichert, Energy-aware Design Space Exploration for GPGPUs, *Computer Science - Research and Development*, 2014, pp. 171-176.

[13]. P. Libuschewski, D. Kaulbars, D. Siedhoff, F. Weichert, H. Müller, C. Wietfeld, P. Marwedel, Multi-Objective Computation Offloading for Mobile Biosensors via LTE, in *Proceedings of the Wireless Mobile Communication and Healthcare Conference (Mobihealth)*, 4, 2014, pp. 226-229.

[14]. P. Libuschewski, P. Marwedel, D. Siedhoff, H. Müller, Multi-Objective Energy-Aware GPGPU Design Space Exploration for Medical or Industrial Applications, in *Proceedings of the IEEE Signal-Image Technology and Internet-Based Systems Conference (SITIS)*, 10, 2014, pp. 637-644.

[15]. J. Lenssen, V. Shpacovitch, F. Weichert, Real-time Virus Size Classification using Surface Plasmon Resonance and Convolutional Neural Networks, in *Proceedings of the Bildverarbeitung für die Medizin Conference (BVM)*, 2017, pp. 98-103.

[16]. V. Shpacovitch, I. Sidorenko, J. Lenssen, V. Temchura, F. Weichert, H. Müller, K. Überla, A. Zybin, A. Schramm, R. Hergenröder, Application of the PAMONO-Sensor for Quantification of Microvesicles and Determination of Nano-Particle Size Distribution, *Sensors*, 17, 2, 2017, pp. 244.

[17]. F. Weichert, M. Gaspar, C. Timm, A. Zybin, E. Gurevich, M. Engel, H. Müller, P. Marwedel, Signal Analysis and Classification for Surface Plasmon Assisted Microscopy of Nanoobjects, *Sensors and Actuators B: Chemical*, 151, 2010, pp. 281-290.

[18]. D. Siedhoff, P. Libuschewski, F. Weichert, A. Zybin, P. Marwedel, H. Müller, Modellierung und Optimierung eines Biosensors zur Detektion viraler Strukturen, *Bildverarbeitung für die Medizin. Lecture Notes in Informatics*, 2014, pp. 108-113.

[19]. D. Siedhoff, H. Fichtenberger, P. Libuschewski, F. Weichert, C. Sohler, H. Müller, Signal/Background Classification of Time Series for Biological Virus Detection, in *Proceedings of the 36th German Conference on Pattern Recognition (GCPR)*, 2014.

[20]. D. Siedhoff, A Parameter-Optimizing Model-Based Approach to the Analysis of Low-SNR Image Sequences for Biological Virus Detection, PhD Thesis, *TU Dortmund University*, 2016.

[21]. O. Neugebauer, P. Libuschewski, M. Engel, H. Müller, P. Marwedel, Plasmon-based Virus Detection on Heterogeneous Embedded Systems,

in *Proceedings of the Workshop on Software & Compilers for Embedded Systems (SCOPES)*, 2015, pp. 48-57.

[22]. D. Siedhoff, A. Zybin, V. Shpacovitch, P. Libuschewski, PAMONO Sensor Data: 200 nm_11Apr13_1. doi: 10.15467/e9ofylrdvk, 2014.

[23]. W. Lorensen, E. Cline, Marching Cubes: A High Resolution 3D Surface Construction Algorithm, in *Proceedings of the 14th Annual Conference on Computer Graphics and Interactive Techniques*, 1987, pp. 163–169.

[24]. L. Breiman, Random Forest, *Machine Learning*, 45, 2001, pp. 5-32.

[25]. T. Hastie, R. Tibshirani, J. Friedman, Unsupervised Learning, The Elements of Statistical Learning: Data Mining, Inference, and Prediction, *Springer*, 2009, pp. 485–585.

[26]. K. Müller, S. Mika, G. Rätsch, K. Tsuda, B. Schölkopf, An introduction to kernel-based learning algorithms, *IEEE Transactions on Neural Networks*, 12, 2001, pp. 181-201.

Chapter 4

MEMS Magnetic Biosensors

Zhen Yang and Fei Wang

4.1. Introduction to MEMS Magnetic Biosensor

4.1.1. MEMS Technology

Micro-Electro-Mechanical Systems (MEMS) is a technology, which can basically be defined as the development of device structures in the microns and even nanometer dimensions using the techniques of microfabrication. Compared with over other types of implantable systems, MEMS technology has some unique advantages for certain applications such as small size scale, electrical nature, and ability to operate on short time scales. During the past decade, MEMS has a rapid development and been used in variety of fields.

The manufacture of MEMS is mainly based on integrated-circuit (IC) technology and micro mechanical processing technology. Compared with traditional microelectronics and machining technology, MEMS technology has several significant features as below: a) Miniaturization. MEMS technology has come to the level of microns and even nanometer. MEMS devices have some advantages such as smaller size, lower power consumption, quicker response, smaller inertia and high portability. b) Integration. Different sensitive direction and braking direction sensors and actuators can be integrated and form sensor array, and even can be integrated with IC for more complex micro system. c) By the silicon as the basic material. Some complex micro 3D structures such as channels,

Zhen Yang

School of Physics and Electronic Engineering, Xinyang Normal University, China pyramids or V-grooves can be obtained, due to the special characteristics of the material silicon. d) Lower cost.

More recently, a large number of MEMS for physics, chemistry, biology and medicine, have been proposed [1].

4.1.2. What are BioMEMS ?

Biomedical MEMS (BioMEMS), a subset of MEMS, are a heavily research area, with a rapid growth in applications of MEMS devices in the biomedical field [2-3]. In general, BioMEMS can be describe as devices or systems used for processing, delivery, manipulation, analysis, and construction of biological and chemical entities, which are constructed using techniques of microfabrication. BioMEMS shows a good prospect in improving the human condition and reducing the cost of health care delivery.

The research field of BioMEMS is extensive, including novel materials for BioMEMS, implantable BioMEMS, micro-fluidics, tissue engineering, surface modification, systems for drug delivery, etc. [4]. Especially, the field of BioMEMS related with bio-technology is very alluring, because it is promising for the miniaturization of the medical diagnosis system.

And with the development of nano-technology, information technology and lab-on-a-chip, the application area of bioMEMS is extending, and it is possible to obtain miniature medical diagnosis system. Therefore, many researchers around the world are performing research on this area.

4.1.3. Magnetic Biosensors

Magnetic sensors with sufficiently high sensitivity are able to detect the bio-magnetic fields produced by the biologic tissues or organs, thus providing a non-invasive mean to detect the activity of the living systems. So, magnetic bio-sensing technology has attracted the global scholars' attention.

The investigation of magnetic biosensor was firstly reported in 1998 by Baselt [5]. The research used a magnetoresistive sensor to detect the presence of micron-sized magnetic particles which can be used as bimolecular labels. Since then, various kinds of magnetic biosensors were proposed for biomagnetic detections, including Fluxgate sensor [6-9], Hall sensor [10-13], Tunnel magnetoresistive sensor (TMR) [14-16], Giant magnetoresistance sensor (GMR) [17-22], Spin valve sensor [23-27], anisotropic magnetoresistance sensor (AMR) [28-33] and Giant magnetoimpedance sensor (GMI) [34-37].

4.2. Fabrication of MEMS-based Magnetic Biosensors

4.2.1. Giant Magnetoimpedance (GMI) Sensors

4.2.1.1. GMI Effect

The giant magnetoimpedance (GMI) effect is a change in the complex impedance at high frequency (usually > 0.1 MHz) and arises from the magnetic field-induced change of the dynamic relative permeability μ_r of the magnetic material, as shown in Fig. 4.1. The effect has been reported in a variety of material geometries, initially in wire, but it has been also observed in glass-coated microwire [38-39], ribbon [40], micro-patterned ribbon [41-42], single layer thin film [43] and multilayer thin film [44].

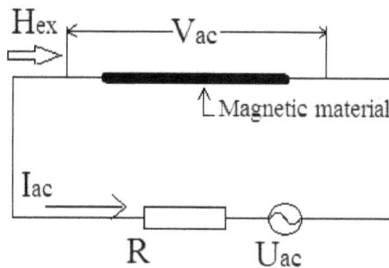

Fig. 4.1. Schematic for the definition of the GMI effect.

GMI has been explained by means of Maxwell equations solved for certain geometries and particular models [45]. According to the theory, it is considered that the physics origin of GMI effect is normally attributed to a combination in the skin depth (δ_m) and high sensitivity of transverse (μ_T) or circumferential permeability (μ_φ) to the external applied field [43, 45]. The ac impedance (Z) of a ferromagnetic ribbon can be calculated by [45]:

$$Z = R_{dc}\, jk\alpha \coth(jk\alpha),$$

where α is the half of the thickness of the ribbon, R_{dc} is the electrical resistance for a dc current, j= imaginary unit, $k = (1+j)/\delta_m$. In a magnetic medium, the δ_m can be given by $\delta_m = (\rho/\pi\mu_T f)$, where ρ is the electrical resistivity, and f is the frequency of the ac current. The dc applied field changes the skin depth through the modification of μ (μ_φ or μ_T) which finally results in a change of the impedance.

4.2.1.2. MEMS Process of GMI Sensor

For a GMI material to be employed for GMI sensor applications, two main requirements should be met, namely, a high GMI ratio (or a large GMI effect) and a high sensitivity to the applied field (or a high magnetic response).The studies in recent years indicated that high GMI effect can be reached in multilayer and meandering thin films. In addition, Among GMI material, Cobalt-based amorphous ribbons with nearly vanishing magnetostriction ($\lambda \sim 0$) have been reported to exhibit GMI effects with a high degree of field sensitivity and are therefore promising candidate materials for making advanced magnetic sensors. In this section, MEMS process of multilayer thin films GMI sensor and micro-patterned ribbon GMI sensors will be introduced.

The multilayer thin films GMI sensor involving the sensing elements of NiFe/Cu/NiFe was fabricated by MEMS technology, including thick photoresist lithography and electroplating. The manufacturing process can be depicted briefly as follows: (a) the 100 nm thick Cr/Cu seed layer was deposited on a glass substrate by radio frequency magnetron sputtering. (b) Photoetching. (c) The bottom NiFe layer was coated by electrodeposition. (d) The Cu layer was accomplished with similar process (e) the top NiFe layer was got. (f) The seed layer was removed by reactive ion etching. The preparation process of high-integrated GMI sensor was shown in Fig. 4.2. The Fig. 4.3 showed the fabricated multilayer thin films GMI sensor.

Cobalt-based amorphous ribbons with meander structure are fabricated by bonding, lithography, chemical etching, and electroplating micro-fabrication process. The manufacturing processes of ribbon GMI biochip platform consisted of the following steps: (a) Attach cobalt-based to Si substrate; (b) Spin photoresist and pattern the ribbon; (c) Etching of the ribbon; (d) Remove the photoresist; (e) Spinning polyimide (PI) on the surface of the sensor and polish; (f) Depositing SiO_2 thin film; (g) Spin photoresist; (h), Sputterthe Cr/Au films. Fig. 4.4 showed the fabrication steps of the Cobalt-based ribbon GMI sensor. Fig. 4.5 showed the fabricated ribbon GMI sensor.

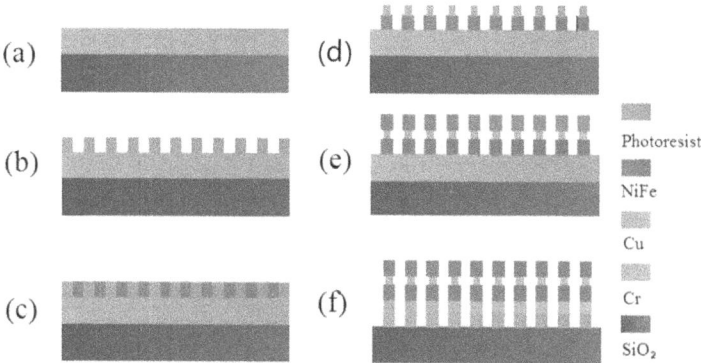

Fig. 4.2. Fabrication steps of the multilayer thin films GMI sensor.

Fig. 4.3. Photograph of the fabricated multilayer thin films GMI sensor.

4.2.1.3. GMI Bio-sensing Principle

As the GMI effect generally occurs at a frequency of ~ MHz order, the surface or the vicinity of the GMI sensor is very sensitive to its magnetic and electric environment due to a strong skin effect. So the GMI bio-sensing principle can be described as below: the surface or the vicinity of the GMI sensor is modified for capturing different biological sample, indirect measurement of biological sample is realized by detection of magnetic particles as the label of biological sample with GMI sensor.

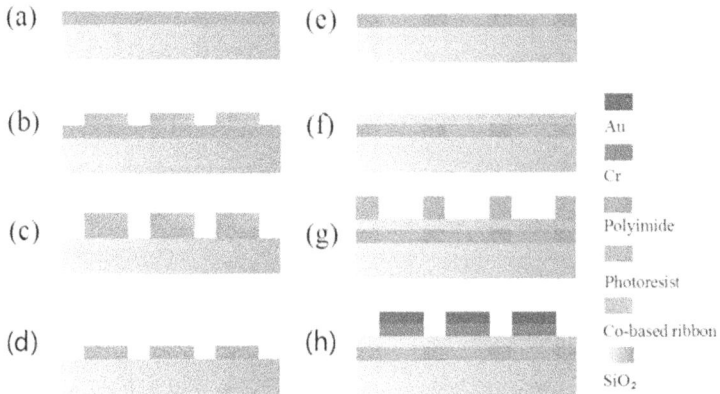

Fig. 4.4. Fabrication steps of the ribbon GMI sensor.

Fig. 4.5. Photograph of the fabricated ribbon GMI sensor.

For detailed expression, when the GMI sensor is subjected to a high frequency alternating current and low frequency external magnetic field, the change in the ac complex impedance $\left(\dfrac{\Delta Z}{Z}\%\right)$ occurred. When the magnetic lables combined with a certain number of biological molecules are present on the surface or near the fringe of the GMI sensor, the magnetic lables becomes magnetized due to the external magnetic field and behaves as a magnetic dipole producing a stray field. Therefore the external magnetic field experience change, and change ratio of the GMI undergo some changes. Finally, we can detect the biomarker qualitatively or quantificationally by considering the difference between the GMI ratios of the GMI sensor with and without magnetic lables. The basic principle diagram is showed in Fig. 4.6.

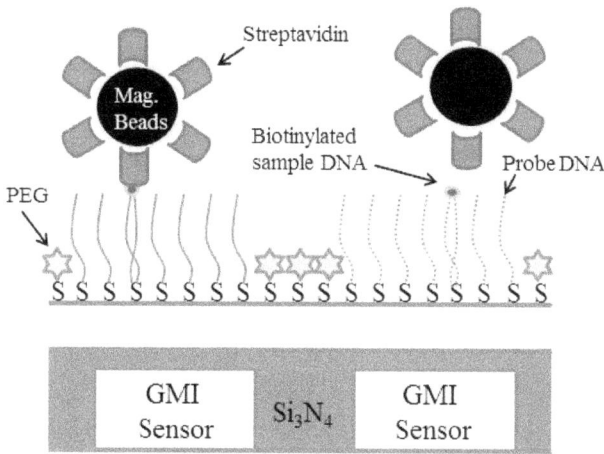

Fig. 4.6. The principle diagram of GMI biosensor.

4.2.2. Giant Magnetoresistance (GMR) Sensors

4.2.2.1. GMR Sensor

Magnetoresistance (MR) sensors are electrical resistors whose resistance changes under the influence of external magnetic fields. This MR effect is caused by the spin–orbit coupling between conduction electrons and magnetic layers. A representative example is the giant magnetoresistance (GMR). The phenomenon takes place in a magnetic structure consisting of ferromagnetic layers separated by non-magnetic metallic layers. The relative change between the layer's magnetization significantly affects the overall electrical resistance.

GMR sensors are also compatible with standard silicon IC technology and suitable for integration into a lab-on-chip system. So the idea of using a GMR in combination with magnetic beads working as magnetic labels for detecting molecular recognition event has received increasing attention.

4.2.2.2. GMR Sensor

The MEMS-based GMR sensor has a top spin valve structure: Si/Ta(5)/seedlayer/IrMn(8)/CoFe(2)/Ru(0.8/CoFe(2)/Cu(2.3)/CoFe(1.5)/ /Ta(3), Each chip sensor consists of 200 strips in serial connection. Each strip has an electrical active area of 300 μm × 3 μm. The photograph of the fabricated GMR sensor is shown in Fig. 4.7. Fig. 4.8 shows the fabrication steps of the GMR sensor. The manufacturing process consisted of the following steps: 1. Spin photoresist and exposure; 2. Pattern spin valve sensors with ion beam etching and remove photoresist; 3. Spin photoresist and exposure; 4. Deposit Cu layer; 5. Lift-off; 6. Spin photoresist and exposure; 7. Deposit SiO2; 8. Lift-off; 9. Spin photoresist and exposure; 10. Sputter Cr/Au; 11. Lift-off.

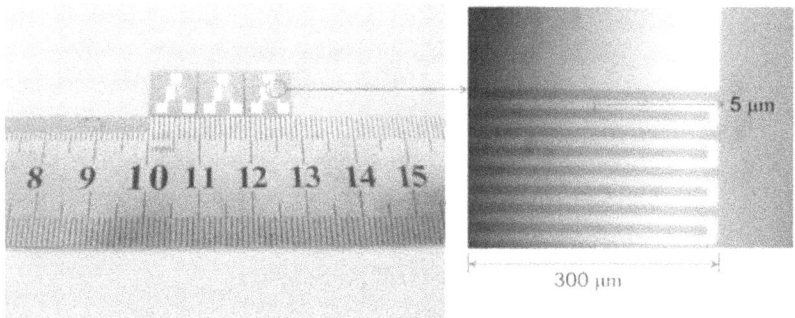

Fig. 4.7. Photograph of the fabricated GMR sensor.

4.2.2.3. GMR Bio-sensing Principle

The working principle of GMR biosensors is similar to that of a GMI biosensor. In GMR biosensor, the stray field of magnetic biomarkers also plays a crucial role in changing the resistance of the GMR sensor. However, their origin for the change is different. GMI effect originates from the change in transverse permeability while GMR effect originates from the change in spin direction of the surface magnetic moments [46].

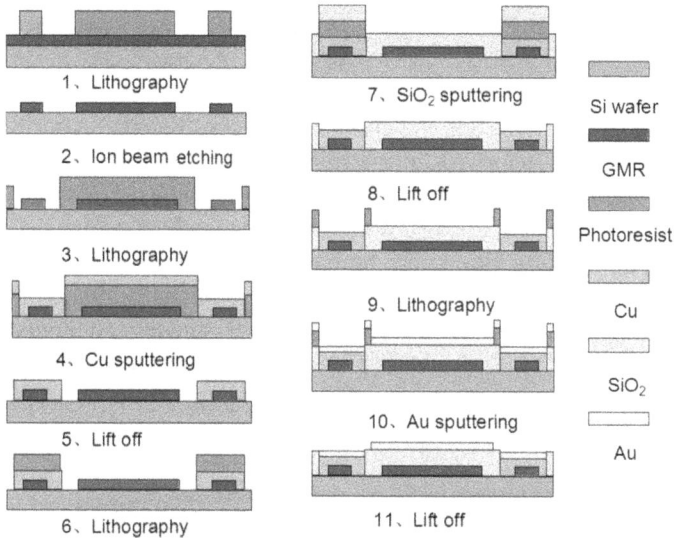

Fig. 4.8. Fabrication steps of the GMR sensor.

4.2.3. Micro Fluxgate Sensors

4.2.3.1. MEMS Fluxgate Sensors

Fluxgate sensor measures DC or low-frequency AC magnetic fields and possesses many advantages over the other magnetic sensors. In terms of resolution of the sensors, fluxgate sensor resolution is higher than the other solid-state devices such as Hall Effect and magnetostrictive sensors, and is comparable to ultrahigh sensitive but very expensive quantum-effect SQUIDs.

Comparing the traditional fluxgate sensors, the fluxgate sensor probes made by MEMS technology have many advantages, such as small sizes, light weight and low consumption. Moreover, this new kind of fluxgate sensors are very suitable to integrate with interface circuits, which greatly decreases the volume and the weight of the whole sensor systems. The MEMS-based method of manufacturing micro fluxgate sensors can reduce the diversity of the sensors in the same production order and the production cost. And it can be used in applications of bio-nanoparticle detection, GPS positioning, nano/pico-satellite attitude control and for game control equipment, etc.

4.2.3.2. Design and Fabrication of MEMS Fluxgate Sensors

In this section, we will introduce the design and fabrication of MEMS fluxgate sensors. In the MEMS fluxgate sensors, the core materials have amorphous ribbon, nanocrystalline ribbon and permalloy. Different core materials were made by different MEMS technologies. The structures of core were also design for different structures such as equal side width rectangle, an unequal side width rectangle, a group of rectangular loops and a helix in Fig. 4.9. 3D solenoid coils were usually made of copper and were fabricated to act as excitation and sensing elements of the sensors. Two excitation coils and a sensing coil in each sensor were placed vertically and wound around different parts of the magnetic core.

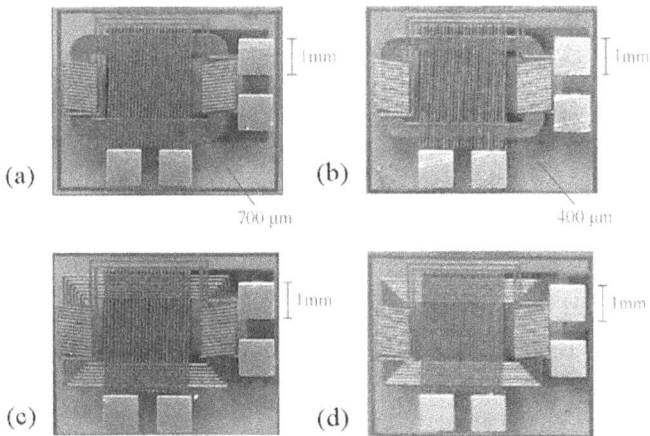

Fig. 4.9. Photographs of the fluxgate sensors with different core structures.

Amorphous and nanocrystalline ribbons core element were fabricated by individual MEMS etching technology and glued on the bottom conductors of MEMS fluxgate. However, the Permalloy was fabricated by electroplating. Take the MEMS fluxgate sensor with permalloy core as an example, the fabrication process was shown in Fig. 4.10 and can be described briefly as follows: 1. The double-side alignment symbols was fabricated on the one side of the glass wafer. A Cr/Cu seed layer of 100 nm was deposited on the other side of the glass wafer. 2. Copper conductors and via were electroplated in the molds. 3. The photoresist and seed layer were removed. 4. Polyimide was spun and solidified on the wafer. 5. A Cr/Cu seed layer was deposited and a photoresist layer was spun on it. 6. The magnetic core was got by electroplating. 7. Cu via was electroplated in the molds and the photoresist and seed layer were removed. 8. Polyimide was spun and a Cr/Cu seed layer was deposited. 9. Top coils were got. 10, the photoresist and seed layer were removed.

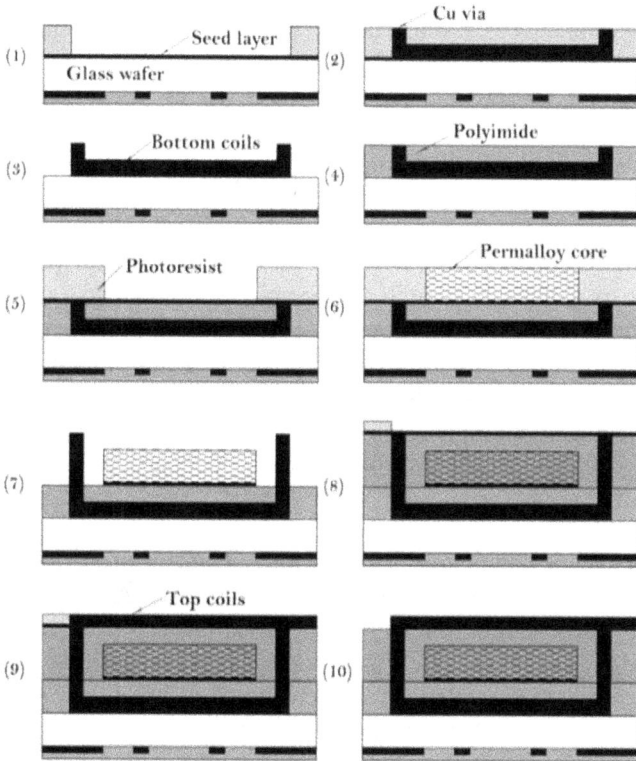

Fig. 4.10. Fabrication steps of the Micro fluxgate sensor.

4.2.3.3. Micro Fluxgate Bio-sensing Principle

The micro fluxgate bio-sensing is also based on detection of magnetic labels. The basic principle on which magnetic bead detection based is that when Dynabeads is placed near the micro fluxgate sensor and exposed to the dc magnetic field, a magnetic dipole is induced in the magnetic beads aligned to the external dc field. The field in the in-plane case is mostly unidirectional and opposite to the external dc field. So, in this case, the presence of magnetic beads can decrease the effective dc magnetic field experienced by the micro fluxgate sensor, and consequently alters the measured results of the sensor.

4.3. Detection of Magnetic Labels

Magnetic nanoparticles (MNPs) labeling is a critical part of magnetic sensing, which converts molecular information into measurable magnetic signals. For sensitive detection, labeling methods aim to maximally load MNPs onto target molecules. Significant Magnetic particle labels and magnetoelectronic detection offer a number of advantages for biosensing. Notably, there is no significant magnetic background present in most samples of interest, thereby enabling detection and magnetic manipulation, both in vitro and in vivo, without affecting the biological interactions.

Microscale magnetic beads have been extensively developed over the past 15 years, primarily for cell and protein separation, and are available from a number of commercial sources. A prime requirement for use as labels is that the beads must be paramagnetic or non-remanent to avoid clustering caused by residual magnetic moment in the absence of magnetic fields. One typical fabrication approach, used for Invitrogen's Dynal M-270, M-280, and M-450 micron-scale beads, is infusion of iron compounds

In recent years, detection of magnetic labels based on magnetic measurement has been presented constantly. A GMI-biosensor fabricated using an amorphous ribbon was employed to detect magnetic Dynabeads [35, 47]. Flexible NiFe/Cu/NiFe multilayered GMI biosensors were presented to detect the Dynabeads protein A and streptavidin-coupled Dynabeads [48-50]. Detection of magnetic-particle concentration in continuous flow based in GMI effect was performed [34]. Detection of the Dynabeads with a MEMS-based micro fluxgate sensor was performed by Zhou et al. [8, 51]. Many researchers have done a lot of magnetic beads testing work by GMR sensors [18, 20].

4.4. Detection of Biomarker Based Magnetic Measurement

4.4.1. GMI Biosensor for Detection of Biomarkers

Wang et al. summarized the magnetoimpedance effect in soft ferromagnetic wires, ribbons and thin films for biosensing applications [52]. Recently, we used a GMI biosensor to detect the Escherichia coli (E. coli) O157:H7, C-reactive protein (CRP) and Myoglobin (Mb) [53-55]. Fig. 4.11 showed the separated-type method based on the GMI biosensor for detection of E. coli O157:H7. In the experiment, Escherichia coli was captured and labeled with Dynabeads by sandwich assay combined with biotin-streptavidin bond on several Au films. Different concentration Escherichia coli (100, 300 and 500 cfu/ml) was first immobilized on Au films, and then placed on the surface of GMI sensor for no-contact testing, respectively. A lower detectable concentration with 50 cfu/ml was achieved. Fig. 4.12 showed in-situ method based on the GMI biosensor for CRP and Mb. This bioassay for CRP and Mb has a linear detection range between 1 and 10 ng/ml. The detection limits were 1 ng/ml and 0.5 ng/ml respectively.

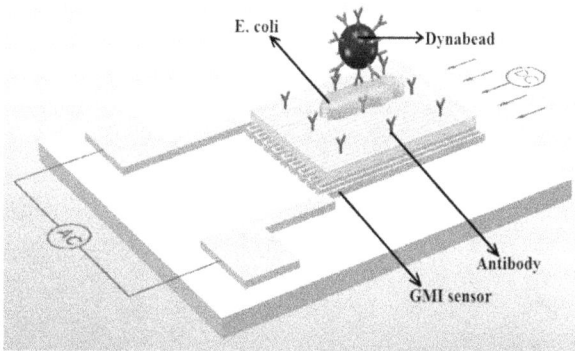

Fig. 4.11. The GMI-based biosensing system for detection of E. coli O157:H7.

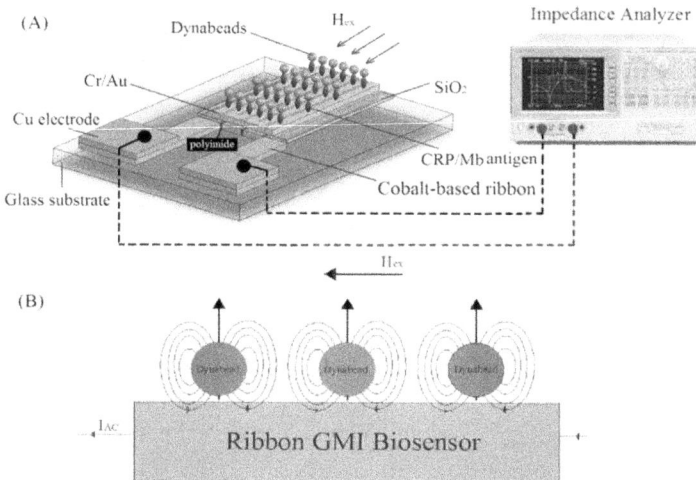

Fig. 4.12. The GMI-based biosensing system for detection of CRP and Mb.

113

4.4.2. GMR Biosensor for Detection of Biomarkers

Sun et al. reported MEMS-based GMR biosensors for detection of E. coli O157:H7, prostate specific antigen (PSA) and carcinoembryonic antigen (CEA) [56-58]. Fig. 4.13 showed the position relationship between GMR sensor and the sample and magnetic field arrangement of the beads under an applied magnetic field. The GMR bio-sensing system was based on double-antibody sandwich assay and streptavidin–biotin binding assay. Dynabeads are used as labels for biomarkers. Under the DC magnetic field, the superparamagnetism Dynabeads induced a low-intensity magnetic field that was detected by GMR sensing element and thus revealing the presence of biomarkers.

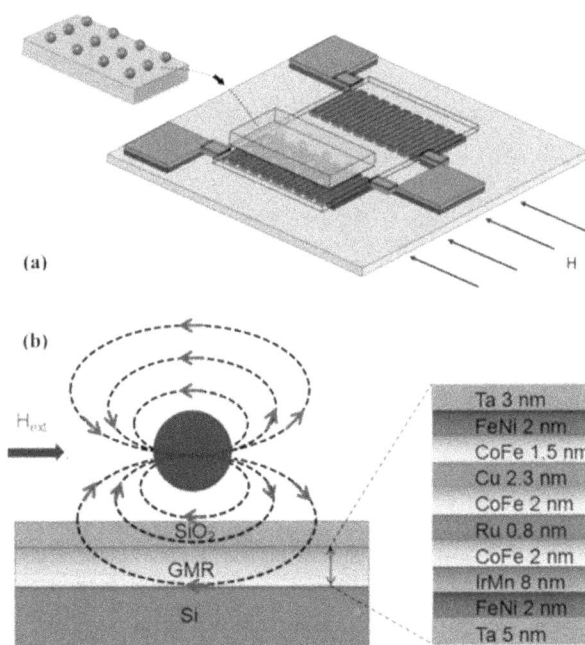

Fig. 4.13. (a) The GMR-based biosensing system.

4.4.3. Micro Fluxgate Biosensor for Detection of Biomarkers

MEMS fluxgate sensors are introduce into the detection of antigens labeled with Dynabeads in recent years. In 2013, Zhou et al. used a micro-fluxgate-based bio-sensing system with Fe-based amorphous core for the detection of alpha fetoprotein (AFP) and CEA [59-60]. A minimum detectable concentration of 1 pg/ml was achieved. Detecting of E. coli O157:H7 by a micro-fluxgate-based bio-sensing system is also achieved [61]. In 2016, a micro fluxgate sensor with rectangular magnetic core was employed for the determination of (PSA) with detection limit as low as 0.1 ng/ml [62]. Fig. 4.14 showed the block diagram and detection discipline of the micro-fluxgate-based biosensing system. Based on the advantages of micro fluxgate sensor, the bio application will certainly expand to other pathogenic bacteria, such as Salmonella, Shigella and Vibrio parahemolyticus, and other biomarkers.

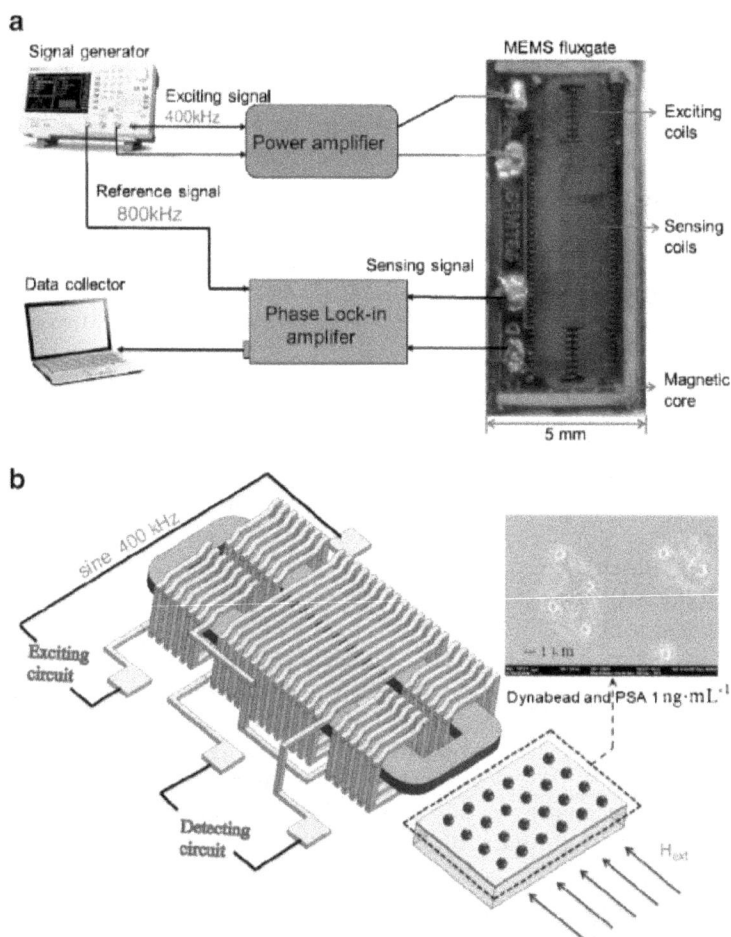

Fig. 4.14. The micro-fluxgate-based bio-sensing system for detection of PSA.

4.5. Conclusion

Benefiting from the inherently negligible magnetic background of biological objects, magnetic detection is highly selective even in complex biological media. MEMS Magnetic sensors have emerged as a powerful diagnostic platform and enables sensitive detection of rare cells and small amounts of molecular markers. Detections of AFP antigens, CEA antigens, CRP antigens, Mb antigen, SPA antigen and E. coli O157:H7 were achieved by using the MEMS magnetic sensor combined with the magnetic labeling technique. We herein summarize recent advances in MEMS magnetic sensors and biomedical applications, with an emphasis on fabrication of MEMS-based GMI, GMR, fluxgate sensor and their applications for detection of biomarkers. MEMS magnetic biosensor is a promising technology for fast and sensitive molecular diagnostics.

References

[1] M. Madou, Fundamentals of Microfabrication: The science of miniaturization, Second Edition, *CRC Press*, 2002.

[2]. S. S. Saliterman, Fundamentals of BioMEMS and Medical Microdevices, *SPIE*, 2006, p. 153.

[3]. B. Ziaie, A. Baldi, M. Lei, Y. Gu, and R. Siegel, Hard and soft micromachining for BioMEMS: Review of techniques and examples of applications in microfluidics and drug delivery, *Adv. Drug Deliv. Rev.,* 56, 2004, 145–172.

[4]. R. Bashir, BioMEMS: State-of-the-art in detection, opportunities and prospects, *Adv. Drug Deliv. Rev.,* Vol. 56, 2004, pp. 1565–1586.

[5]. D. R. Baselt, G. U. Lee, M. Natesan, S. W. Metzger, P. E. Sheehan, R. J. Colton, A biosensor based on magnetoresistance technology, *Biosens. Bioelectron,* 1998, 13, pp. 731–739.

[6]. F. Ludwig, E. Heim, S. Mäuselein, D. Eberbeck, M. Schilling, Magnetorelaxometry of magnetic nanoparticles with fluxgate magnetometers for the analysis of biological targets, *J. Magn. Magn Mater,* 293, 1, 2005, pp. 690–695.

[7]. F. Ludwig, S. Mäuselein, E. Heim, M. Schilling, Magnetorelaxometry of magnetic nanoparticles in magnetically unshielded environment utilizing a differential fluxgate arrangement, *Rev. Sci. Instrum,* 76, 10, 2005, p. 106102.

[8]. J, Lei, T. Wang, C. Lei, Y. Zhou, Detection of targeted carcinoembryonic antigens using a micro-fluxgate-based biosensor, *Appl Phys Lett*, 102, 2013, p. 022413.

[9]. E. Heim, F. Ludwig, M. Schilling, Binding assays with streptavidin-functionalized superparamagnetic nanoparticles and biotinylated analytes using fluxgate magnetorelaxometry, *J. Magn. Magn Mater,* 321, 2009, pp. 1628–1631.

[10]. P. A. Besse, G. Boero, M. Demierre, V. Pott, R. Popovic, Detection of a single magnetic microbead using a miniaturized silicon Hall sensor, *Appl. Phys. Lett,* 80, 22, 2002, pp. 4199–4201.

[11]. W. Lee, S. J. Joo, U. K. Sun, K. W. Rhie, J. K. Hong, K. H. Shin, K. H. Kim, Magnetic bead counter using a micro-Hall sensor for biological applications, *Appl. Phys. Lett,* 94, 2009, 153903.

[12]. B. Sinha, T. S. Ramulu, K. W. Kim, R. Venu, J. J. Lee, C. G. Kim, Planar Hall magnetoresistive aptasensor for thrombin detection, *Biosens. Bioelectron,* 59, 2014, pp. 140–144.

[13]. A. Sandhu, Y. Kumagai, A. Lapicki, S. Sakamoto, M. Abe, H. Handa, High efficiency Hall effect micro-biosensor platform for detection of magnetically labeled biomolecules, *Biosens. Bioelectron,* 22, 9, 2007, pp. 2115–2120.

[14]. C. Albon, A. Weddemann, A. Auge, K. Rott, A. Hütten, Tunneling magnetoresistance sensors for high resolutive particle detection, *Appl. Phys. Lett,* 95, 2, 2009, 023101.

[15]. W. Shen, X. Liu, D. Mazumdar, G. Xiao, In situ detection of single micron-sized magnetic beads using magnetic tunnel junction sensors, *Appl. Phys. Lett.,* 86, 25, 2005, 253901.

[16]. F. Li, J. Kosel, An efficient biosensor made of an electromagnetic trap and a magneto-resistive sensor, *Biosens. Bioelectron.,* 59, 2014, pp. 145–150.

[17]. A. Shoshi, J. Schotter, P. Schroeder, M. Milnera, P. Ertl, R. Heer, H. Brueckl, Magnetic lab-on-a-chip for cell analysis: magnetoresistive-based real-time monitoring of dynamic cell-environment interactions, *Biosens. Bioelectron.,* 2013, 40, 1, pp. 82–88.

[18]. B. Srinivasan, Y. Li, Y. Jing, Y. Xu, X. Yao, C. Xing, J. P. Wang, A Detection System Based on Giant Magnetoresistive Sensors and High-Moment Magnetic Nanoparticles Demonstrates Zeptomole Sensitivity: Potential for Personalized Medicine, *Angew. Chem. Int. Ed. Engl.,* 48, 15, 2009, pp. 2764–2767.

[19]. M. Mujika, S. Arana, E. Castano, M. Tijero, R. Vilares, J. M. Ruano-Lopez, J. Berganza, Magnetoresistive immunosensor for the detection of Escherichia coli O157:H7 including a microfluidic network, *Biosens. Bioelectron,* 24, 5, 2009, pp. 1253–1258.

[20]. R. S. Gaster, D. A. Hall, C. H. Nielsen, S. J. Osterfeld, H. Yu, K. E. Mach, R. J. Wilson, B. Murmann, J. C. Liao, S. S. Gambhir, S. X. Wang, Matrix-insensitive protein assays push the limits of biosensors in medicine, *Nat. Med.,* 2009, 15, 11, pp. 1327–1332.

[21]. J. Choi, A. W. Gani, D. J. Bechstein, J. R. Lee, P. J. Utz, S. X. Wang, Portable, one-step, and rapid GMR biosensor platform with smartphone interface, *Biosens. Bioelectron,* 85, 2016, pp. 1–7.

[22]. B. M. De Boer, J. A. H. M. Kahlman, T. P. G. H. Jansen, H. Duric, J. Veen, An integrated and sensitive detection platform for magneto-resistive biosensors, *Biosens. Bioelectron,* 22, 9, 2007, pp. 2366–2370.

[23]. D. Serrate, J. M. De Teresa, C. Marquina, J. Marzo, D. Saurel, F. A. Cardoso, M. R. Ibarra, Quantitative biomolecular sensing station based on magnetoresistive patterned arrays, *Biosens. Bioelectron,* 35, 1, 2012, pp. 206–212.

[24]. V. C. Martins, F. A. Cardoso, J. Germano, S. Cardoso, L. Sousa, M. Piedade, L. P. Fonseca, Femtomolar Limit of Detection with a Magneto-resistive Biochip, *Biosens. Bioelectron,* 24, 8, 2009, pp. 2690–2695.

[25]. H. A. Ferreira, D. L. Graham, P. P. Freitas, J. M. S. Cabral, Biodetection using magnetically labeled biomolecules and arrays of spin valve sensors, *J. Appl. Phys.,* 93, 2002, pp. 7281–7286.

[26]. D. L. Graham, H. Ferreira, J. Bernardo, P. P. Freitas, J. M. S. Cabral, Single magnetic microsphere placement and detection on-chip using current line designs with integrated spin valve sensors: Biotechnological applications, *J. Appl. Phys.,* 91, 10, 2002, pp. 7786–7788.

[27]. D. L. Graham, H. A. Ferreira, P. P. Freitas, J. M. S. Cabral, High sensitivity detection of molecular recognition using magnetically labelled biomolecules and magnetoresistive sensors, *Biosens. Bioelectron,* 18, 4, 2003, pp. 483–488.

[28]. S. Oh, M. Jadhav, J. Lim, V. Reddy, C. Kim, An organic substrate based magnetoresistive sensor for rapid bacteria detection, *Biosens Bioelectron,* 41, 2013, pp. 758–763.

[29]. P. A. Besse, G. Boero, M. Demierre, V. Pott, R. Popovic, Detection of a single magnetic microbead using a miniaturized silicon Hall sensor, *Applied Physics Letters,* 80, 2002, pp. 4199–4201.

[30]. G. Rizzi, F. W. Østerberg, M. Dufva, M. F. Hansen, Magnetoresistive sensor for real-time single nucleotide polymorphism genotyping, *Biosens Bioelectron,* 52, 2014, pp. 445–451.

[31]. M. M. Miller, G. A. Prinz, S. F. Cheng, S. Bounnak, Detection of a micron-sized magnetic sphere using a ring-shaped anisotropic magnetoresistance-based sensor: A model for a magnetoresistance-based biosensor, *Appl. Phys. Lett.,* 2002, 81, 12, pp. 2211–2213.

[32]. Z. Jiang, J. Llandro, T. Mitrelias, J. A. C. Bland, Enabling suspension-based biochemical assays with digital magnetic microtags, *J. Appl. Phys.,* 99, 8, 2006, pp. 08S105.

[33]. R. S. Gaster, L. Xu, S. J. Han, R. J. Wilson, D. A. Hall, S. J. Osterfeld, H. Yu, S. X. Wang, Quantification of Protein Interactions and Solution Transport Using High-Density GMR Sensor Arrays, *Nature Nanotechnology,* 6, 2011, pp. 314–320.

[34]. G. V. Kurlyandskaya, M. L. Sanchez, B. Hernando, V. M. Prida, P. Gorria, M. Tejedor. Giant-magnetoimpedance based sensitive element as a model for biosensors, *Appl Phys Lett,* 82, 2003, pp. 3053–3055.

[35]. G. V. Kurlyandskaya, V. Levit, Magnetic Dynabeads detection by sensitive element based on giant magnetoimpedance, *Biosen Bioelectron*, 20, 8, 2005, pp. 1611-1616.

[36]. G. V. Kurlyandskaya, Giant magnetoimpedance for biosensing: advantages and shortcomings, *J.Magn Mater*, 321, 2009, p. 659.

[37]. H. Chiriac, M. Tibu, A. E. Moga, Magnetic GMI sensor for detection of biomolecules, *J. Magn. Magn Mater*, 293, 1, 2005, pp. 671-676.

[38]. J. P. Sinnecker, M. Nobel, K. R. Pirota, J. M. Garcia, A. Asenjo, and M. Vazquez, Frequency dependence of the magnetoimpedance in amorphous CoP electrodeposited layers, *J. Appl. Phys.*, 87, 2000, p. 4825.

[39]. M. Ipatov, V. Zhukova, A. Zhukov, Gozález, Magnetotransport at High Frequency of Soft Magnetic Amorphous Ribbons, *Appl. Phys. Lett.*, 97, 2010, 252507.

[40]. F. Alves, L. Abi Rached, J. Moutoussamy, C. Coillot, Trilayer GMI sensors based on fast stress-annealing of FeSiBCuNb ribbons, *Sens. Actuators A*, 142, 2008, 459.

[41]. L. Chen, Y. Zhou, C. Lei, Z.M. Zhou, Giant magnetoimpedance effect and voltage response in meander shape Co-based ribbon, *Appl. Phys. A.*, 98, 2010, p. 861.

[42]. Z. Yang, J. Lei, C. Lei, Y. Zhou, T. Wang, Effect of magnetic field annealing and size on the giant magnetoimpedance in micro-patterned Co-based ribbon with a meander structure, *Appl. Phys. A*, 116, 2014, pp. 1847-1851.

[43]. R. L. Sommer and C. L. Chine, Asymmetric magnetoimpedance in two-phase ferromagnetic film structures, *Appl. Phys. Lett*, 47, 1995, p. 3346.

[44]. T. Wang, Z. Yang, C. Lei, J. Lei, Y. Zhou, An integrated giant magnetoimpedance biosensor for detection of biomarker, *Biosens. Bioelectron*, 58, 2014, pp. 338–344.

[45]. M. H. Phan, H. X. Peng., Giant magnetoimpedance materials: Fundamentals and applications, *Progre. Materi Sci*, 53, 2008, p. 323.

[46]. M. N. Baibich, J. M. Broto, A. Fert, F. N. Vandau, F. Petroff, P. Eitenne, G. Creuzet, A. Friederich, and J. Chazelas, Giant Magnetoresistance of (001)Fe/(001)Cr Magnetic Superlattices, *Physical Review Letters*, 61, 1988, 2472.

[47]. Z. Yang, C. Lei, Y. Zhou, Y. Liu, X. Sun, A GMI biosensing platform based on Co-based amorphous ribbon for detection of magnetic Dynabeads, *Anal. Methods*, 7, 16, 2015, pp. 6883–6889.

[48]. T. Wang, Y. Zhou, C. Lei, J. Lei and Z. Yang, Development of an ingenious method for determination of Dynabeads protein A based on a giant magnetoimpedance sensor, *Sens Actuators B.*, 186, 2013, pp. 727-733.

[49]. T. Wang, Z. Yang, C. Lei, J. Lei and Y. Zhou, A giant magnetoimpedance sensor for sensitive detection of streptavidin-coupled Dynabeads, *Phys. Status Solidi A*, 211, 6, 2014, pp. 1389-1394.

[50]. T. Wang, Y. Zhou, C. Lei, J. Lei and Z. Yang, Ultrasensitive detection of Dynabeads protein A using the giant magnetoimpedance effect, *Microchim Acta.,* 180, 2013, pp. 1211-1216.

[51]. J. Lei, C. Lei, T. Wang, Z. Yang, Y. Zhou, A MEMS-fluxgate-based sensing system for the detection of Dynabeads, *J. Micromech. Microeng,* 23, 2013, 095005.

[52]. T. Wang, Y. Zhou, C. Lei, Jun. Luo, Shaorong Xie and Huayan Pu, Magnetic impedance biosensor: A review, *Biosens. Bioelectron,* 90, 2017, pp. 418-435.

[53]. Z. Yang, Y. Liu, C. Lei, X. Sun, Y. Zhou, A flexible giant magnetoimpedance-based biosensor for the determination of the biomarker C-reactive protein, *Microchim. Acta,* 182, 2015, pp. 2411–2417.

[54]. Z. Yang, X. Sun, T. Wang, C. Lei, Y. Liu, Y. Zhou, J. Lei, A giant magnetoimpedance-based biosensor for sensitive detection of Escherichia coli O157:H7, *Biomed. Microdevices,* 17, 1, 2015, pp. 1–8.

[55]. Z. Yang, Y. Liu, X. Sun, Y. Zhou, Ultrasensitive detection and quantification of E. coli O157:H7 using a giant magnetoimpedance sensor in an open-surface microfluidic cavity covered with an antibody-modified gold surface, *Microchim. Acta,* 183, 2016, pp. 1831–1837.

[56]. X. Sun, S. Zhi, C. Lei, Y. Zhou, Investigation of contactless detection using a giant magnetoresistance sensor for detecting prostate specific antigen, *Biomed. Microdevices,* 183, 2016, pp. 1107-1114.

[57]. X. Sun, C. Lei, L. Guo, Y. Zhou, Giant magneto-resistance based immunoassay for the tumor marker carcinoembryonic antigen, *Microchim. Acta,* 183, 2016, pp. 2385-2393.

[58]. X. Sun, C. Lei, L. Guo, Y. Zhou, Separable detecting of Escherichia coli O157H:H7 by a giant magneto-resistance-based bio-sensing system, *Sens Actuators B.,* 234, 2016, pp. 485-492.

[59]. J. Lei, C. Lei, T. Wang, Z. Yang, Y. Zhou, Investigation of targeted biomolecules in a micro-fluxgate-based bio-sensing system, *Biomed. Microdevices,* 16, 2014, pp. 237-243.

[60]. J. Lei, T. Wang, C. Lei, Y. Zhou, Detection of targeted carcinoembryonic antigens using a micro-fluxgate-based biosensor, *Appl. Phys. Lett,* 102, 2013, 022413.

[61]. X. Sun, Z. Yang, C. Lei, Y. Liu, L. Guo, Y. Zhou, An innovative detecting way of Escherichia coli O157H:H7 by a micro-fluxgate-based bio-sensing system, *Sensors and Actuators B: Chemical,* 221, 2015, pp. 985-992.

[62]. X. Sun, C. Lei, L. Guo, Y. Zhou, Sandwich immunoassay for the prostate specific antigen using a micro-fluxgate and magnetic bead labels, *Microchim. Acta,* 183, 2016, pp. 2385-239.

Chapter 5

Application of Surface Photo-Charge Effect for Control of Fluids

José Luis Pérez-Díaz and Mariana K. Kuneva

5.1. Introduction

The surface photo-charge effect (SPCE) has been observed as a new type of electromagnetic field-matter interaction during studies on TAV (transverse acoustoelectric voltage) effect [1-3]. TAV is induced by the interaction of a solid sample situated above a piezoelectric with the electromagnetic field alongside an acoustic wave propagating in the piezoelectric. During studies on TAV, results were obtained that were initially inexplicable [4-6]. It was established that SPCE is a specific feature of each solid; during the interaction of a solid with a modulated electromagnetic field, the solid generates an alternating electric signal with a frequency equal to that of the modulation [7, 8]. SPCE could be used in many practical applications for solids characterization as visualization of implanted regions of solids [9], detection of mechanical imperfections on semiconductor/metal surfaces, inhomogeneities in the impurities/dopants distribution [10, 11], study of electric properties (type of conductivity) [12] and electron states [13-15] of semiconductor samples, visualization of electron surface topography [16], express contactless chemical composition test of samples [17], monitoring the quality of raw materials [18, 19].

The SPCE is very sensitive to the state of the irradiated surface and thus could give information about fluids themselves and the processes taking place in them as well. The review presented reveals our experimental

J. L. Pérez-Díaz
Universidad Carlos III de Madrid, Instituto Pedro Juan de Lastanosa, Spain

setups and results on application of SPCE for studying of fluids and demonstrates how SPCE could be used for study of fluids on some examples including: liquid identification [20], monitoring of the octane factor of gasoline [21], impurities in fluids (fluid composition) [21], monitoring of material deposition from solution [21], liquid level meter [22], milk and food control [23].

5.2. SPCE

The SPCE takes place when any solid surface is being illuminated with intensity-modulated electromagnetic field and manifests itself in generating an alternating current (AC) between the solid and common ground across the light direction [9, 10, 24]. The amplitude of the signal created by SPCE depends on the specific characteristics of the irradiated surface but the AC frequency remains the same as that of the light modulation. The SPCE is a universal property of any kind of solid [25], and has been observed in a very wide range of the electromagnetic spectrum [26]. A principal scheme of the SPCE observation is shown in Fig. 5.1.

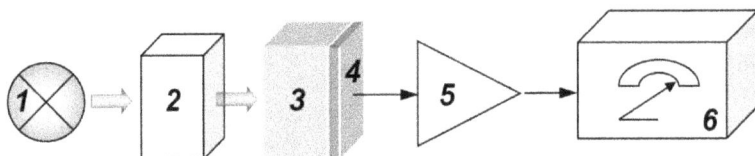

Fig. 5.1. A general scheme of the SPCE observation: 1 – light source; 2 – modulator; 3 – irradiated solid; 4 – measuring electrode; 5 – amplifier; 6 – signal registering device.

Two possible models have been discussed for explaining SPCE in the case of conductors [10] and in the case of dielectrics [24]. The first suggests generation of force perpendicular to the illuminated surface when incident light attenuates in depth of the solid and redistribution of the charges in the conductor takes place. When the solid is a dielectric one, SPCE could be explained as due to changes in the charge of surface energy states at the irradiated surface. The possibility of SPCE to be due to a photodesorption or surface sputtering induced by the incident radiation has been discussed in a series of articles [27-29]. The model completely explaining the SPCE still needs some clarifying.

5.3. SPCE at the Liquid-Solid Interface

As it has been observed [21], any change in the electron properties of the solid-liquid interface results in change of the SPCE signal obtained when the interface is being irradiated by intensity-modulated light. The suggestion that the SPCE in liquids is not an inherent property of the liquid itself but is rather generated at the solid–liquid interface was confirmed by the results of investigations of opaque liquids, when no signal could be measured with sufficient experimental accuracy unless this interface was irradiated.

A scheme of the measuring structure for SPCE in liquid-solid interface is shown in Fig. 5.2. When an irradiated solid surface (S) is in contact with a fluid being investigated (F), any kind of changes in the fluid properties could be detected by the change of the SPCE signal since the surface states of the solid/liquid interface (I) depend on the contacting fluid as well. The SPCE voltage is measured on the electrode (E), coupled to suitable equipment. In the case of gas or vapor sensors, a solid with maximal adsorption capacity to the respective fluid should be used as a substrate.

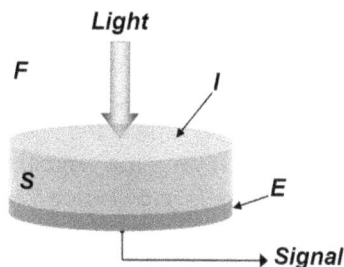

Fig. 5.2. Scheme of the measuring structure for SPCE in liquid-solid interface: S-solid; I-solid-liquid interface generating the signal; F-fluid studied; E – electrode.

To increase the amplitude of the detected signal, the illuminated substrate has to be very sensitive to SPCE - as semiconductors are - and monochromatic light is preferable. As an optimizing step, the wavelength of the illuminating light has to be adjusted too, to obtain an optimal response for each investigated fluid. Also, the light intensity, the modulating frequency range, the method for signal detection, the parameters of the measuring structure should be selected for each case of fluid in such a way that a minimal change in the controlled variable will cause maximal changes in the measured electrical signal. Optimization is necessary to increase the sensitivity and to avoid possible hysteresis, slow relaxation etc., for each particular case of fluid monitoring.

As it has been mentioned above, SPCE is a feature of each solid illuminated by intensity-modulated light. The presence of a liquid most probably changes the capacity of the measuring structure at the illuminated solid-liquid interface, so an increase in the SPCE signal is observed [21].

During our experiments described below, changes in the SPCE-signals were observed strongly depending on variations in the fluid composition and properties (temperature, volume, velocity, pressure, etc.). It is clear that the processes taking place at the interface depend strongly on the liquid characteristics but at present their contribution to the SPCE signal cannot be separated from the possible contribution of different processes responsible for the generation of the SPCE voltage in each particular case. For example, in a structure with direct contact between the conducting electrodes and the liquid, electro-chemical reactions can play an important role.

It is very important to note that the signals were obtained not only from conducting electrodes directly contacting the liquid, but from electrodes placed out of the liquid as well, since the method is capacitive. In contactless configuration however, the signal is weaker and the measurement requires special tools (e.g. Teflon vessel or glazed ceramic crucible).

5.4. Applications

Since the SPCE signal is highly sensitive to any changes in the composition and the properties of the studied fluid, this effect is very attractive for prospective practical applications as a complementary, rapid analysis together with the already existing methods for liquid control. Under certain conditions, the SPCE signal is a function of the composition or other properties of the liquid and could be implemented as an analytical method, similar to the potentiometric methods in analytical chemistry. The first study of the SPCE signal at liquid-solid interface was reported in 2000 [20] with an emphasis on its application for fluid (gas and vapor) sensors.

It has to be underlined that the SPCE-based method is universal and could be used for any fluid under various conditions. It is contactless, rapid and the results are measured in real time. Most of the equipment for its implementation is off-the-shelf.

5.4.1. SPCE-Signal for Liquid Identification

A setup for SPCE detection is shown in Fig. 5.3 [21]. A He-Ne laser (at $\lambda=633$ nm, and 50 mW) or an Ar-ion laser (at $\lambda=488$ nm, and 250 mW) was used as light source (1). The illumination spot was the same in all experiments in order to avoid setup effects. When the studied liquid is in a vessel (scheme (a)) the SPCE signal comes from the electrodes (4) placed in or out of the vessel. The signal magnitude depends on the form, material and size of the electrodes and the vessel and their configuration as well as on the light spot location. Fifteen milliliters of liquid were used in each measurement with the same spot location. If a drop of liquid is measured, a substrate (Teflon or semiconductor one) (1) acts as a vessel and the SPCE signal comes from the electrode (4) and the grounded metal box (5) used for avoiding electrical noises.

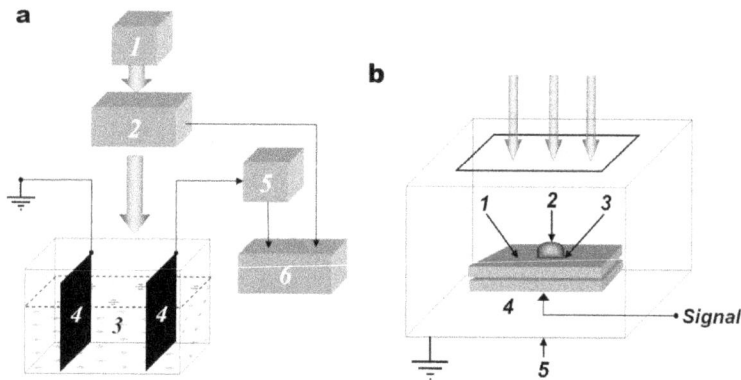

Fig. 5.3. Experimental scheme for SPCE detection with electrodes immersed into the liquid (a): 1 - light source; 2 - modulator; 3 - vessel with liquid; 4 - electrodes for signal measurement; 5 - pre-amplifier; 6 - lock-in nanovoltmeter. Configuration for SPCE measurement with a drop of liquid (b): 1 - teflon or semiconductor substrate; 2 - liquid drop; 3 - liquid–solid interface; 4 - electrode; 5 - metal grounded box.

A characteristic feature of the SPCE in liquids is that the potential difference created by this effect has a specific value for each liquid with substantial difference between them. For example, the amplitudes of the measured signal for several, randomly selected liquids when an Ar-laser was used as light source are: (a) - tap water - 190 μV; (b) - filtered water - 209 μV; (c) - distilled water - 175 μV; (d) - alcohol – 115 μV; (e) - ammonia solution - 65 μV; (f) - acetone - 50 μV; (g) - coffee (used as an opaque liquid) - 15 μV. The results are schematically presented in Fig. 5.4.

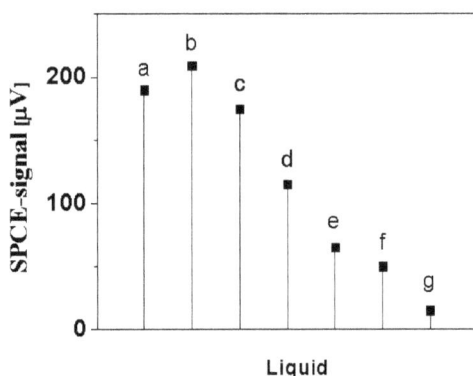

Fig. 5.4. The SPCE signal for different liquids described in the text.

The results presented above demonstrate that the SPCE signal considerably varies from liquid to liquid and that it is also possible to detect qualitatively variations in some liquid characteristics. The data for three types of water (filtered, distilled and from the tap) show that even insignificant treatment of the liquid causes easily detectable signal variations. Thus, analysis and control of drinking water can be carried out.

A quantitative analysis of any changes in the liquid characteristics could also be possible, but it requires a different type of experiments.

5.4.2. Monitoring of the Octane Factor of Gasoline

As an example of the fact that various modifications of a given liquid could be detected by changes in the SPCE signal, measurements were performed for gasoline samples having different octane grades. The measurements were carried out under the same experimental conditions as described above with the exception that the amount of liquid was 30 ml and the liquid vessel was partially covered by a metal cover kept for all measurements. That cover was placed in the same position in order to keep its effect on the signal amplitude unchanged. The gasoline samples were supplied by the Bulgarian Center for Standardization. A quality control device could be developed since the SPCE signal differs for leaded or unleaded gasoline with various octane grades: (a) - A-86 - 29 μV; (b) - A-91 /unleaded/ -59 μV; (c) - A-93 - 54μV; (d) - A- 93/unleaded/ - 63 μV; (e) - A-95 /unleaded/ - 64 μV; (f) - A-96 - 46 μV. The amplitudes of the SPCE signal for a series of fuels, purchased from regular gas stations were: diesel - 48 μV; A-86 - 31 μV; A-93 - 52 μV; A-95 /unleaded/ - 68 μV. The results are schematically presented in Fig. 5.5.

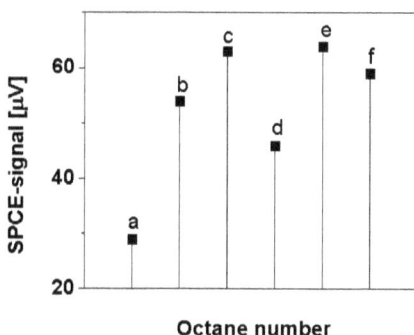

Fig. 5.5. The SPCE signal for samples of different octane number as described in the text.

The experiments described above demonstrate in principle the possibility for qualitative detection of different liquids or variations in some liquid characteristics when all other experimental parameters are kept constant. Further development and calibration of the proposed method would make possible a quantitative analysis as well.

5.4.3. Detection of Impurities in Liquids

Remarkable variations of the SPCE signal were also found upon mixing two liquids [21]. Some experiments were performed in a glass vessel by using an Ar-laser as an illuminating source. The results showed that if only one drop of ammonia solution was added to 200 ml of drinking water, a 15 % decrease in the amplitude of the signal was detected, whereas the signal increased by 10 % when one drop of acetone was added in the same amount of water.

Thus, any change in the chemical composition, due to contamination with noxious substances, can be detected by the corresponding variation in the SPCE signal. Also, various types of absorbing filters for gases and liquids can be monitored in order to determine when they have to be replaced. This is possible due to the fact that the absorption (of the filtered substances) changes the chemical composition of the filter. The latter can be detected by SPCE.

SPCE can be used to monitor the environmental pollution as well. The test is non-destructive, continuous and real-time and would find application in areas such as ecology, industry, car production, military equipment, etc.

5.4.4. Monitoring of Material Deposition from a Solution

Our experiments have shown that the precipitation of a substance from a solution on the volume surface could also be monitored by the SPCE-method. Since a deposition modifies the surface, it changes the SPCE signal. Such a change, for example, was observed during $CaCO_3$ deposition from water solution on a metal surface [21], the signal amplitude being dependent on the intensity of the illuminating light.

5.4.5. Level Meter

Another group of experiments was performed showing that by varying the type of electrode and the illumination conditions, the SPCE signal may become dependent on the level of the liquid. This provided an opportunity for another device – a new type of level meter for liquids - to be designed.

A laboratory model of a level meter for liquid fuel (unleaded 95 H for example) was developed. Its principal sketch is shown in Fig. 5.6: (1) is a solid, shaped as a stick with a measuring structure deposited on it. Its length is larger or at least equal to the height of the container where the liquid (3) is stored. The stick was irradiated with modulated electromagnetic field of ultra-high frequency (UHF) instead of illuminating with light. The SPCE signal was measured with an electrode (2) which is also placed along the length of the stick. When the level changes it causes a change in the area of the exposed solid surface or in the area of solid/liquid interface. In the level meter case, the irradiated surface is in contact with the fluid to be controlled.

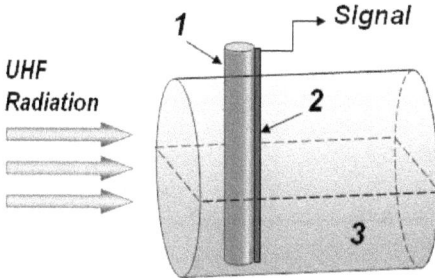

Fig. 5.6. Scheme of a level meter: 1 - solid stick with deposited measuring structure on it; 2 – electrode; 3 – liquid.

One of the problems to be solved in any sensor based on the SPCE is the best choice of material of the solid, accuracy, depth of the tank (size of the solid surface) and working conditions for each particular case of liquid of interest. When the level meter is properly built and optimized, the signal could be very sensitive to small changes (of the order of 0.1 mm) in the level of the liquid, and proportionality between the SPCE signal and the level of the liquid in the tank is observed (Fig. 5.7).

As far as the measurement of liquid fuel levels is the most widespread case, a small laboratory model was built for testing unleaded petrol 95H. About 100 % change of the electric signal was observed when the change of the liquid level was from 0 to 10 cm. The measurement accuracy was 0.4 mm and it can be increased simply by technical improvement. The device makes also possible to measure the level of the water usually being present at the bottom of the tank by simultaneous measurement of the fuel level.

Although the investigations performed so far did not reveal any principal difficulties for the implementation of such device, some problems affecting the measurement accuracy have to be solved before the device leaves its laboratory-prototype stage. The condensation of vapors on the measuring structure, for example, could decrease the accuracy of measurement and lead to errors. A way to avoid such problems is to improve the measuring structure in order to minimize the influence of the condensation, or to put the condensation under control. Some preliminary tests in this respect were carried out. Hot water vapor was supplied to a closed chamber, in which the level meter had been placed, until saturation was achieved. At these extreme conditions, the experimental error was about 1 mm.

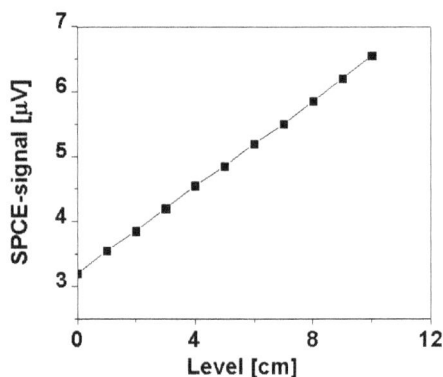

Fig. 5.7. SPCE response of the laboratory level meter versus the petrol level.

Since the generated potential differences are very small - of the order of nanovolts and microvolts - the measuring technique does not pose risk of sparks. In addition, the device can be designed with electrical wiring separated from the liquid fuel.

Devices based on SPCE are competitive to most of the popular technical solutions available currently. The level meter described above has the advantages of a simple design, small investments for production, sensitivity to very small variations in the liquid level, measurement in real time and no moving parts being involved. The flowing velocity of a liquid can also be monitored by SPCE, since, at certain conditions, the generated voltage depends on this factor as well.

5.4.6. Milk Quality Control

A large number of analytical methods have been developed for milk investigation, including biochemical, microbiological, serological methods, etc. Their principal disadvantages are related to their slowness, problems with consumables, the need of laboratory conditions as well as to the features of the reagents required - some of them are expensive, extremely unstable, toxic or carcinogenic. Instantaneous tests, not requiring any consumables, are still in great demand.

The experiments with SPCE showed that it is possible to detect some specific changes and processes taking place in the milk [23, 24] offering a technology for rapid detection of inhibitors in milk.

The scheme of the experimental setup is similar to the one shown in Fig. 5.2. The light source (1) was a continuous wave diode laser generating 25 mW at λ=655 nm whose beam was chopped by a modulator (2) into periodic pulses with a modulation frequency of 800 Hz (far from the one of the electric network - 50 Hz – to avoid possible parasitic signals). A pulsed laser or a pulsed LED could also be used instead of a mechanical modulator. The milk sample (10 ml) was placed in a small vessel in which two electrodes are placed, but a setup capable to perform the test on only a drop of liquid is also possible. A high-impedance pre-amplifier (5) with a gain of 20 dB at input resistance of 10^8 Ω and a lock-in nanovoltmeter (6) were used for detecting the SPCE-signals, the reference signal to the lock-in being supplied by the modulator.

The measured signal is formed at the milk–solid interface, so any minor changes in the milk (concentration, contaminations, pretreatment, etc.) change the SPCE signal as discussed above. The required information for milk quality can be obtained either by direct observation of the signal amplitude or by using appropriate liquid reagents to modify the signal. The milk samples were characterized by measuring the amplitude of the signal generated by pure milk and its variation after a droplet (0.04 ml) of testing liquid was added to the milk. The testing liquid (kanamycin) was introduced in both the incident spot of the laser beam and at a certain distance (1–3 cm) away from it. The results presented in Fig. 5.8 clearly show that the signal strongly depends on the nature of the liquid (inhibitor: antibiotic, preservative or water) added to the milk sample. At the same time, the dependence of the signal on the laser beam spot location underlines the importance of varying only the liquid properties with all other experimental conditions being kept constant.

An opportunity to detect changes in milk due to the presence of exogenous substances is demonstrated in Figs. 5.8 and 5.9. The arrow signs indicate the moment of time at which the testing liquid was dripped in the milk sample. Fig. 5.10 shows the change in the signal when droplets of acetic acid are being added to the milk one by one. The SPCE response was faster and signal amplitude (145 µV for pure milk) changed the strongest after the first droplet. Some saturation effect (weak and time-delayed signal) was observed which could be due to the increased acidity of the milk sample when the response to the test liquid decreases.

Variations in the signal response, similar to those presented in Fig. 5.8, were also observed in the presence of inhibitors in the milk samples (Fig. 5.9) when the effect of adding 0.04 ml of kanamycin was followed in samples of pure milk and milk containing hydrogen peroxide. The signal amplitude of the milk sample with inhibitor (30 % solution of H_2O_2) was 753 µV (Fig. 5.8 (a)) and changed to 162 µV after adding kanamycin (Fig. 5.9 (b)).

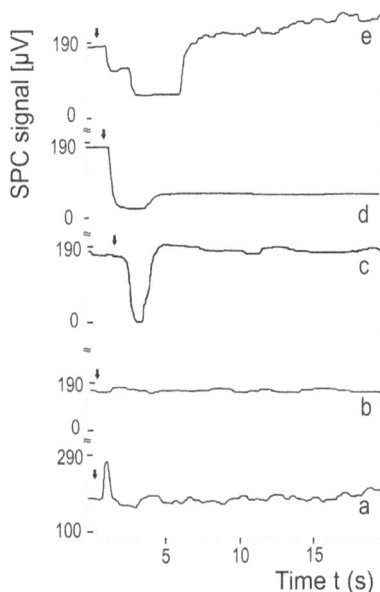

Fig. 5.8. Effect of various testing liquids on the SPCE-signal amplitude: a) Antibiotic - pharmazin (added in the incident spot of the laser beam); b) Water (added in the incident spot of the laser beam); c) Concentrated acetic acid (added 1 cm away from the incident spot of the laser beam); d) 30 % solution of a preservative - H_2O_2 (added 1 cm away from the incident spot of the laser beam); e) Antibiotic - tetravet (added in the incident spot of the laser beam).

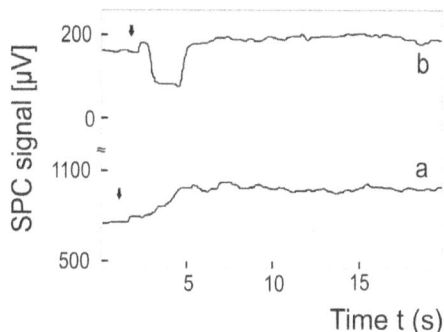

Fig. 5.9. Influence of the antibiotic kanamycin dripped in the incident spot of the laser beam on the signal amplitude for 20 s: a) After addition of 10 drops of H_2O_2 to 200 ml of milk (being stored 2.5 hours before the test); b) Signal of pure milk.

Although the shape of the curve in Fig. 5.9 (a) depended on the period of time, during which the hydrogen peroxide was present in the milk, it always remained different from the curve in Fig. 5.9 (b). This allows a method for instantaneous tests to be developed for qualitative characterization of the milk in respect to the presence of hydrogen peroxide in it without any need of substantial expenses for consumables. Furthermore, to detect inhibitors, it is not even necessary to record the signal, but only to compare its amplitude with the amplitude of pure milk signal: that of the pure milk was 4.46 times lower than the one of the sample containing H_2O_2.

It has to be mentioned that since the shape of the curve in Fig. 5.9 (a) depends on how long the test liquid (H_2O_2) has been present in the milk, it suggests also a possibility to study the processes taking place in milk due to the presence of hydrogen peroxide.

The SPCE-signal is very sensitive to any minor change in the milk (concentration, contaminations, pretreatment, etc.) as could be seen in Fig. 5.10. Beside the opportunity for quality control of milk, the method described above could be used for recognition of milk samples produced by different animal species since they generate signals with different amplitudes. The following types of milk were analyzed: cow, sheep and buffalo. The amplitude of the signal generated by sheep milk was reduced by 15 % compared to the signal generated by cow milk. The amplitude of the signal generated by the buffalo milk sample was 50 % lower than the one generated by cow milk.

133

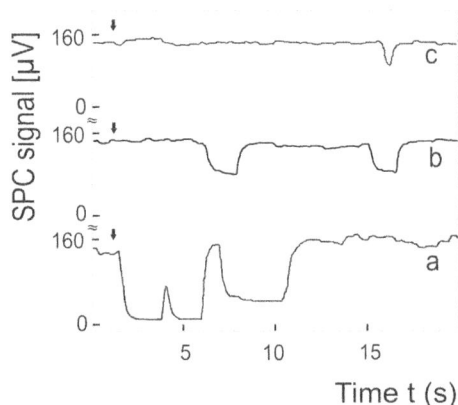

Fig. 5.10. Signal variation for 20 s after dripping concentrated acetic acid in the incident spot of the laser beam: a) After the first drop; b) After the second drop; c) After the third drop. The milk was well-stirred after each drop for homogenization.

Since at the present stage of work the investigations are mostly qualitative ones, they cannot identify the admixture type and its quantity in the milk. Considerable work is necessary to make possible quantitative estimations. Nevertheless, a qualitative evaluation of milk purity could be very useful for practical applications, if performed in a fast way, in real working environment, and without considerable expenses which are the features typical for the techniques proposed here. A device based on SPCE could be made portable, easily operated without expensive consumables, providing fast results, and not requiring laboratory environment.

Beside milk, SPCE can be applied for characterization of other foods (e.g. bee honey) as well.

5.5. Conclusions

The effect presented here has the great potential to complement the already known and used methods for control of fluids with a rapid and contactless method. The SPCE provides a universal testing method since a signal is generated by all types of fluids, and the method could be used for characterization of any fluid. It is possible to monitor even small quantities (a drop) with a suitable measurement setup. The method can easily be implemented in commercial environments, which allows various practical applications to be developed.

The main advantages of SPCE-based methods include: high accuracy, low cost, instantaneous results, and possibilities for rapid field measurements in real time, with no need of complex equipment or qualified personnel, and no metal contact which has to be deposited on the sample. No power supplies are required and the measuring technique does not pose any additional risks (health hazards, fire, defects, etc.). The practical solutions for various cases are simple and do not include large components which allows the development of small-sized sensors. This analytical method gives also great opportunities because it combines optical probing of the sample with electrical detection of the generated signal. Since the SPCE signal changes with different changes in the studied fluid, the sensors could be built for various parameters of interest. The SPCE-based sensors can be used for on-line and/or in-situ control.

Our study on the application of the SPCE for quality control of different materials and processes corresponds to the priorities of the contemporary scientific research focused on the improvement of the quality of life.

References

[1]. O. Ermolova, O. Ivanov, O. Naidov-Zhelezov, I. Solodow, Transverse acoustoelectric effect with bulk wave reflection in piezoelectric-semiconductor structures, *Acoustics Letters*, Vol. 9, Issue 3, 1985, pp. 41- 44.
[2]. L. Konstantinov, V. Strashilov, O. Ivanov, Kinetics and polarity of the transverse acoustoelectric effect in the separated-medium surface acoustic wave configuration, *J. Phys. D: Appl. Phys.,* Vol. 18, Issue 7, 1985, pp. L79 – L85.
[3]. V. Strashilov, L. Konstantinov, O. Ivanov, Topographical studies of semiconductor surfaces by using a combined photo-acoustoelectric method, *Appl. Phys. B*, Vol. 43, Issue 1, 1987, pp. 17- 21.
[4]. M. Borissov, O. Ivanov, V. Kovachev, V. Lovchinov, Transverse acoustoelectric effect in piezoelectric configuration - metal conductive sample*, Acoustics Letters*, Vol. 11, Issue 12, 1988, pp. 229-232.
[5]. O. Ivanov, A review of the major investigations and problems associated with the transverse acoustoelectric effect, *Acoustics Letters*, Vol. 13, Issue 4, 1989, pp. 56-63.
[6]. V. Pustovoit, O. Ivanov, Surface charge redistribution effect in a conductor subjected to electromagnetic radiation, *Comptes Rendus de l'Academie Bulgare des Sciences*, Vol. 42, Issue 4, 1989, pp. 39-42.

[7]. O. Ivanov, Sensor applications of field-matter interactions, in Encyclopedia of Sensors, Grimes C. A., Dickey E. C. & Pishko M. V. (Eds.), *American Scientific Publishers,* Stevenson Ranch, California, Vol. 9, 2006, pp. 165-197.

[8]. O. Ivanov, V. Mihailov, V. Pustovoit, A. Abbate, P. Das, Surface photo-charge effect in solids, *Optics Communications*, Vol. 113, Issue 1, 1995, pp. 509-512.

[9]. V. Pustovoit, M. Borissov, O. Ivanov, Photon charge effect in conductors, *Physics Letters A,* Vol. 135, Issue 1, 1989, pp. 59-61.

[10]. V. Pustovoit, M. Borissov, O. Ivanov, Surface photo-charge effect in conductors, *Solid State Commun.,* Vol. 72, Issue 6, 1989, pp. 613–619.

[11]. V. I. Pustovoit, M. Borissov, O. Ivanov, Surface photon-charge effect in conductors, *Bulg .J. Physics*, Vol. 17, Issue 1, 1990, pp. 32 – 40.

[12]. P. Das, V. Mihailov, O. Ivanov, V. Georgiev, S. Andreev, V. Pustovoit, Contactless characterization of semiconductor devices using surface photo-effect, *IEEE Electron Device Letters,* Vol. 13, Issue 5, 1992, pp. 291- 293.

[13]. A. Abbate, P. Rencibia, O. Ivanov, G. Masini, F. Palma, P. Das, Contactless characterization of semiconductor, using laser-induced surface photo-charge voltage measurements, *Materials Science Forum,* Vol. 173-174, 1995, pp. 221- 226.

[14]. I. Davydov, O. Ivanov, D. Svircov, G. Georgiev, A. Odrinsky, V. Pustovoit, Contactless spectroscopy of deep levels in semiconducting materials: GaAs, *Spectroscopy Letters,* Vol. 27, Issue 10, 1994, pp. 1281-1288.

[15]. O. Ivanov, L. Konstantinov, Temperature dependence of the surface photo-charge effect in CdS, *Applied Surface Science*, Vol. 143, Issues 1-4, 1999, pp. 101-103.

[16]. O. Ivanov, D. Svircov, Ts. Mihailova, P. Nikolov, V. Pustovoit, Automatized system for measuring the surface density of current carriers and electrical permittivity of conducting materials, *Spectroscopy Letters*, Vol. 28, Issue 7, 1995, pp. 1085-1094.

[17]. O. Ivanov, A. Vaseashta, L. Stoichev, Rapid, contactless, and non-destructive testing of chemical composition of samples, in Functionalized Nanoscale Materials, Devices and Systems for Chem.-bio Sensors, Photonics, and Energy Generation and Storage, Vaseashta, Ashok K., Mihailescu, Ion N. (Eds.), *Springer*, 2008, pp. 331-334.

[18]. O. Ivanov, Zh. Stoyanov, B. Stoyanov, M. Nadoliisky, A. Vaseashta, Fast, contactless monitoring of the chemical composition of raw materials, in Technological Innovations in Sensing and Detection of Chemical, Biological, Radiological, Nuclear Threats and Ecological Terrorism, A. Vaseashta, E. Braman, Ph. Susmann, (Eds.), *Springer*, 2012, pp. 185 - 189.

[19]. O. Ivanov, A. Vaseashta, A method for fast and contactless control of raw materials, *Ceramics International,* Vol. 39, Issue 3, 2012, pp. 2903-2907.

136

[20]. O. Ivanov, L. Konstantinov, Application of the photo-induced charge effect to study liquids and gases, *Surface Review and Letters*, Vol. 7, Issue 3, 2000, pp. 211-212.

[21]. O. Ivanov, L. Konstantinov, Investigation of liquids by photo-induced charge effect at solid–liquid interfaces, *Sens Actuator B,* Vol. 86, Issues 2-3, 2002, pp. 287–289.

[22]. O. Ivanov, Level-meter for liquids based on the surface photo-charge effect, *Sens Actuator* B, Vol. 75, Issue 3, 2001, pp. 210–212.

[23]. O. Ivanov, S. Radanski, Application of surface photo charge effect for milk quality Control, *J Food Sci.,* Vol. 74, Issue 7, 2009, pp. R79–R83.

[24]. O. Ivanov, V. Mihailov, V. Pustovoit, P. Das, Surface photo-charge effect in dielectrics, *Comptes Rendus de l'Academie Bulgare des Sciences,* Vol. 47, Issue 6, 1994, pp. 21-24.

[25]. N. Vancova, O. Ivanov, I. Yordanova, Experimental investigations of surface photo-charge effect in different materials, *Spectroscopy Letters,* Vol. 30, Issue 2, 1997, pp. 257-266.

[26]. O. Ivanov, V. Mihailov, R. Djulgerova, Spectral dependencies of the surface photocharge effect at conducting surfaces, *Spectroscopy Letters*, Vol. 33, Issue 3, 2000, pp. 393-398.

[27]. O. Ivanov, R. Dyulgerova and M. Georgiev, Photoinduced electrification of solids. I. Plausible mechanisms, *xxx.lanl.gov,* Paper ID: cond-mat/0508457, 2005, 17 pgs.

[28]. M. Georgiev and O. Ivanov, Photoinduced electrification of solids. II. Photovoltage transients, *xxx.lanl.gov*, Paper ID: cond-mat/0508460, 2005, 21 pgs.

[29]. O. Ivanov, E. Leyarovski, V. Lovchinov, Chr. Popov, M. Kamenova, M. Georgiev, Photoinduced electrification of solids. III. Temperature dependences, *xxx.lanl.gov*, Paper ID: cond-mat/ 0706.3877, 2007, 10 pgs.

Chapter 6

Electronic Nose

Ozge Cihanbegendi Sahin

6.1. Introduction

Olfaction is one of the major senses of human being. It stimulates our feelings; increases the sense of tasting; prevents us from danger. Artificial olfaction concept arose from the need to sense some odors electronically. Many scientists have been working on developing electronic systems that can mimic the human sense of smell for years.

Although the concept of an electronic nose using a chemical sensor array for classifying odors was proposed by Persaud [1] in 1982, the primary artificial olfaction device is realized by Moncrieeff [2] depending on the absorption of different odorants on various films in 1961. Wilkens, Hatman [3], and Buck [4] developed the first electronic nose in 1964. The term "electronic nose" was first used at the 8th International Congress of European Chemoreception Research Organisation, University of Warwick in 1988 [5]. So, electronic nose concept has been developing for more than fifty years [6]. After 1990s, more extended researches are carried out; the term "artificial/electronic nose" started to be used and several commercial instruments are developed. Especially in the past decade, electronic nose instrumentation has grown internationally because of its potential to solve various problems in fragrance and cosmetics production, food and beverages manufacturing, chemical engineering, environmental monitoring, and more recently, medical diagnostics and bioprocesses.

Ozge Cihanbegendi Sahin
Dokuz Eylul University, Department of Electrical and Electronics Engineering,
Izmir, Turkey

In this chapter, physiological properties of human nose are summarized to be able to see the similarities with an electronic nose. Than the main parts and the features of electronic nose system are explained. Sensors are the essential building blocks of the system that must be selected according to the application area and the chemical gases to be sensed. Also, application areas are introduced and related sensors of some are mentioned.

6.2. Physiology of Human Nose

Human nose includes olfactory epithelium, olfactory bulb and cleft to sense odors as shown in Fig. 6.1.

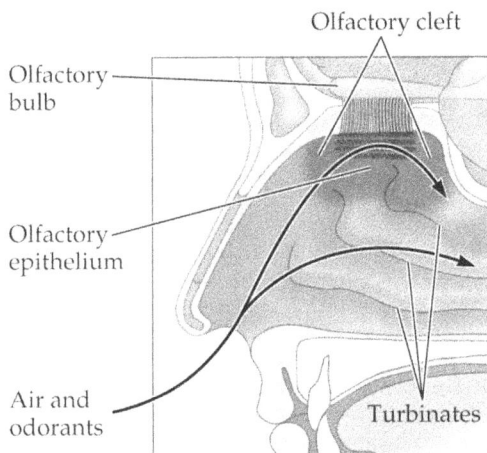

Fig. 6.1. Physiology of human nose [9].

Some terms related to smelling function need to be defined before describing the mechanism:

olfaction: The sense of smell.

odor: The translation of a chemical stimulus into a smell sensation.

odorant: A molecule that is defined by its physicochemical characteristics, which are capable of being translated by the nervous system into the perception of a smell.

olfactory cleft: A narrow space at the back of the nose into which air flows and where the olfactory epithelium is located.

olfactory epithelium: A secretory mucous membrane in the human nose whose primary function is to detect odorants in inhaled air. Located on both sides of the upper portion of the nasal cavity and the olfactory clefts, the olfactory epithelium contains three types of cells: olfactory sensory neurons, basal cells, and supporting cells.

olfactory bulb: A blueberry-sized extension of the brain just above the nose, where olfactory information is first processed. There are two olfactory bulbs, one in each brain hemisphere, corresponding to the right and left nostrils.

The three basic elements, namely the olfactory receptor cells in the olfactory epithelium, the olfactory bulb and the brain form the basis for the development of the artificial olfaction devices.

6.3. E-nose

Electronic nose, shortly "e-nose" is defined by Gardner and Bartlett as an instrument consisting of an array of electronic chemical sensors with special properties and a pattern-recognition system which can recognize simple or complex odors [7]. In fact, this is not so similar to the human nose and Mielle *et al.* stated that such an analytical system is 'obviously electronic but not nose' [8].

In 1980s, by using an array of gas sensors and pattern recognition techniques, E-nose was invented for distinguishing a variety of odours [10].

The need of an e-nose results from the following causes:

- The human sniffers are costly when compared to electronic nose.

- Speedy, reliable new technology of the gas sensors are used in the electronic nose.

- Detection of hazardous or poisonous gas is not possible with a human sniffer.

- An e-nose also overcomes other problems associated with the human olfactory system. • There lies a great difference in the values got by each individual.

Electronic nose consists of a sensing system and PARC (PAttern ReCognition system. Sensing system can be built by an array of several different sensing elements or a single device or a combination of both depending on the odor to be sensed. VOCs (Volatile Organic Components) poses a signature, a fingerprint or a pattern which presents the characteristic of the odor. A data base can be built up by presenting various vapors to the sensor system. PARC methods can be classified in three groups:

- Graphical analysis is a simple form of data analysis.

- Multivariate analysis generally involves data reduction. It reduces high dimensionality in a multivariate problem where variables are partly correlated and allows the information to be displayed in a smaller dimension. Principal component analysis (PCA), Cluster Analysis (CLA), Linear Discriminant Analysis (LDA), Partial Least Squares(PLS) methods are some of these analysis methods.

- Artificial Neural Networks (ANN): A neural network consists of a set of interconnected processing algorithms functioning in parallel. On a very simplified and abstract level, ANN is based on the cognitive process of the human brain.

The selection of the method depends on existing data and the type of result which is required [12].

E-nose originates from human olfaction mechanism. Fig. 6.2 shows the similarity between them.

Both of them include a receptor unit at the input, a processing part to extract the features and a pattern classification unit in the end. E-nose and human nose (bio-nose) can be compared as given in Table 6.1.

6.4. Sensors Employed in E-nose Systems

While many prototype sensors are being produced in research institutions, several commercial sensors are available on the market. These sensors exhibit physical and chemical interactions with the chemical compounds.

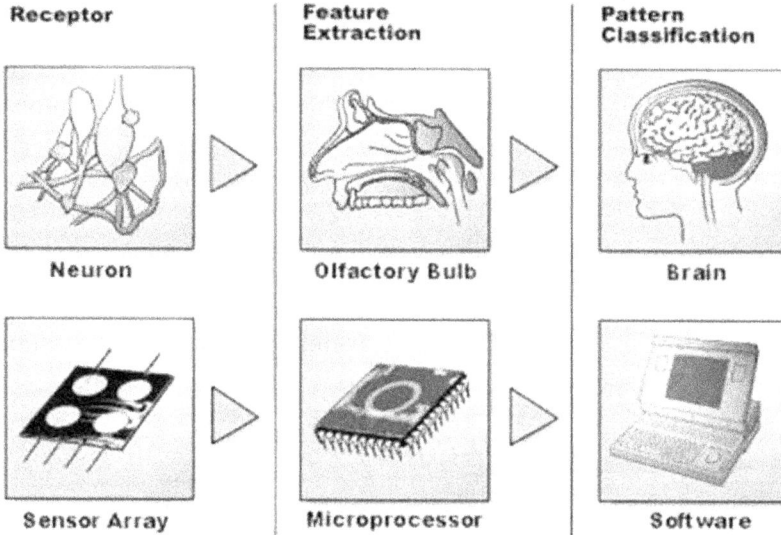

Fig. 6.2. Similarity between human nose and e-nose [11].

Table 6.1. Comparison of e-nose with bio-nose [7].

Bio-nose	Electronic nose
It uses the lungs to bring the odor to epithelium layer	It employs a pump to smell the odor
It has mucus, membrane, and hair to act as filter	It has an inlet sampling system that provides filtration
The human nose contains the olfactory epithelium, which contains millions of sensing cells that interact with odorants in unique	Electronic nose has a variety of sensors that interact differently with a group of odorous molecules
The human receptors convert the chemical response to electronic nerve impulses whose unique patterns are propagated by neurons through a complex network before reaching the higher brain for interpretation	Similarly, the chemical sensors in the electronic nose react with the sample and produce electrical signals. A computer reads the unique pattern of signals and interprets them with some form of intelligent pattern classification algorithms

Sensor responses depend on several variables:

- Flow of the carrier gas;
- Nature of the odour;
- Interaction kinetics;
- Sensing material (nature and substrate);

- Odour diffusion within the sensing material;
- Ambient conditions (temperature, humidity, pressure, etc.) [13].

Gas molecules interact with solid-state sensors by absorption, adsorption or chemical reactions with thin or thick films of the sensor material. The sensor device detects the physical and/or chemical changes incurred by these processes and these changes are measured as an electrical signal. The most common types of changes utilized in e-nose sensor systems are shown in Table 6.2 along with the classes of sensor devices used to detect these changes. Some aspects of these classes of sensors, along with several examples from each class, are explored in the following sections of this review.

Table 6.2. Physical changes in the sensor used to transduce them into electrical signals [4].

Sensor	Physical property
Metal-oxide sensors (MOS, MOSFET)	Resistance and impedance
Conducting polymer sensors	Resistance and impedance
Electrochemical sensors	Conductance, intensity and voltage
Acoustic sensors (SAW, BAW, QMB, Cantilever)	Mass and frequency shift
Calorimetric sensors	Temperature
Optical sensors	Optical properties

Various kinds of gas sensors are available, but only four technologies are currently used in commercialized electronic noses (Fig. 6.3): Metal Oxide Semiconductors (MOS); Metal Oxide Semiconductor Field Effect Transistors (MOSFET); conducting organic polymers (CP); piezoelectric crystals (bulk acoustic wave: BAW and Surface acoustic wave: SAW). Such sensors can be divided into two main classes: hot (MOS, MOSFET) and cold (CP, SAW, BAW). The former operate at high temperatures and are considered to be less sensitive to moisture with less carryover from one measurement to another [14]. MOS and CP sensors are also grouped in Conductivity sensors.

Fibreoptic, electrochemical and bi-metal sensors are still in the developmental stage and may be integrated in the next generation of electronic noses.

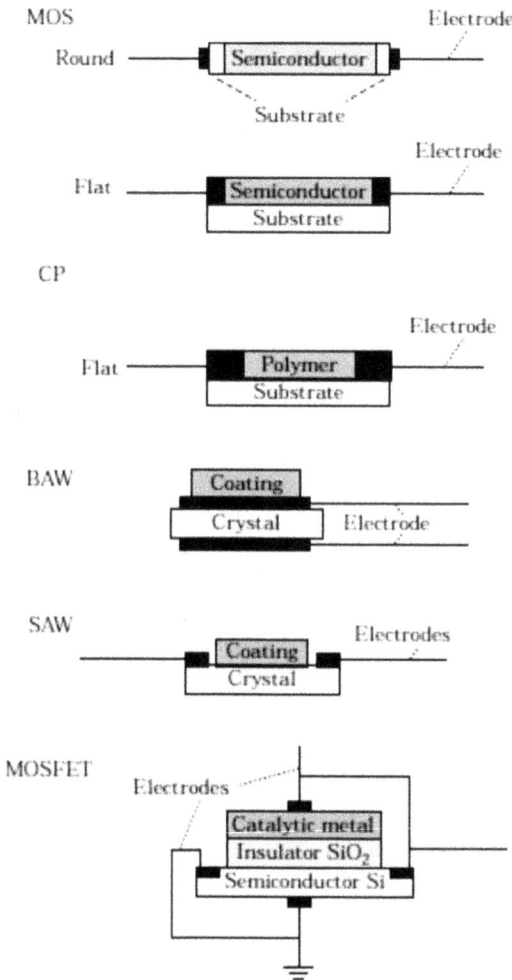

Fig. 6.3. Schematic diagrams of five different kinds of sensors [14].

6.4.1. Metal Oxide Semiconductor (MOS) Sensors

Gas sensors, based on the chemical sensitivity of Metal Oxide Semiconductors (MOS), are commercially available. They have been more widely used to make arrays for odor measurement than any other class of gas sensors. MOS sensors are extremely sensitive to ethanol, which 'blinds' them to any other volatile compound [14].

The sensor usually comprises a ceramic support tube containing a platinum heater coil onto which sintered SnO_2 is coated onto the outside of the tube with any catalytic additives. Gas samples are sensed by the change in the electrical resistance of the metal oxide semiconductor. Resistance changes due to combustion reactions occurring within the lattice oxygen species on the surface of metal oxide particles. Although the oxides of many metals show gas sensitivity under suitable conditions, the most widely used material is tin dioxide (SnO_2) doped with a small amount of a catalytic metal such as palladium or platinum. By changing the choice of catalyst and operating conditions, tin dioxide resistive sensors have been developed for a range of applications [12].

6.4.2. Metal Oxide Semiconductor Field-Effect Transistor (MOSFET) Sensors

The metal oxide semiconductor field-effect transistor (MOSFET) sensors rely on a change of electrostatic potential. A MOSFET sensor consists of three layers, a silicon semiconductor, a silicon oxide insulator and a catalytic metal (usually palladium, platinum, iridium or rhodium), also called the gate. In the MOSFET transistor, the gate and drain contacts are shortcut, giving a diode mode transistor with convenient electronics for operation, characterized by an I-V curve. The applied voltage on the gate and drain contact creates an electric field, which influences the conductivity of the transistor. When polar compounds interact with this metal gate, the electric field, and thus the current flowing through the sensor, are modified. The recorded response corresponds to the change of voltage necessary to keep a constant preset drain current

The selectivity and sensitivity of MOSFET sensors may be influenced by the operating temperature (50–200 °C), the composition of the metal gate, and the microstructure of the catalytic metal. MOSFET sensors, like MOS sensors, have a relatively low sensitivity to moisture and are thought to be very robust. However, high levels of manufacturing expertise are necessary to achieve good quality and reproducibility [14].

6.4.3. Conducting Polymer (CP) Sensors

Conducting organic polymer (CP) sensors have been under development for approximately 10 years and, like MOS sensors, rely on changes of resistance by the adsorption of gas. However, their operating mechanism is more complex and not yet well understood. These sensors comprise a substrate (e.g. fibre-glass or silicon), a pair of gold-plated electrodes, and a conducting organic polymer such as polypyrrole, polyaniline or polythiophene as a sensing element.

In general, these sensors show good sensitivities, especially for polar compounds. However, their low operating temperature (< 50 °C) makes them extremely sensitive to moisture. Although such sensors are resistant to poisoning, they have a lifetime of only about 9–18 months. This short life may be due to the oxidation of the polymer, or to exposure of the sensor to different chemicals that may develop contact resistances between the polymer and the electrodes.

Unlike MOS sensors, the CP sensors are not yet widely available in markets, and laboratory-scale manufacturing renders them expensive. The difficulty of producing good batch-to-batch reproducibility and a pronounced drift of the response are their main disadvantages [14].

6.4.4. Piezoelectric Crystal Sensors (Acoustic Sensors)

Piezoelectric sensors are based on a change of mass, which may be measured as a change in resonance frequency.

These sensors are made of tiny discs, usually quartz, lithium niobate ($LiNbO_3$) or lithium tantalate ($LiTaO_3$), coated with materials such as chromato-graphic stationary phases, lipids or any non-volatile compounds that are chemically and thermally stable.

When an alternating electrical potential is applied at room temperature, the crystal vibrates at a very stable frequency, defined by its mechanical properties. Upon exposure to a vapour, the coating adsorbs certain molecules, which increases the mass of the sensing layer and hence decreases the resonance frequency of the crystal. This change may be monitored and related to the volatile present.

The crystals may be made to vibrate in a Bulk Acoustic Wave (BAW) or in a Surface Acoustic Wave (SAW) mode by selecting the appropriate combination of crystal cut and type of electrode configuration. BAW and SAW sensors have different structures. BAW are 3-dimensional waves travelling through the crystal, and SAW are 2-dimensional waves.

Since piezoelectric sensors may be coated with an unlimited number of materials, they present the best selectivity. However, the coating technology is not yet well controlled, which induces poor batch-to-batch reproducibility. SAW sensors, though limited by the noise caused by their high operating frequency, are more sensitive than BAW sensors. However, both sensors require a higher concentration of volatiles to elicit response levels comparable to other sensor types.

The difficulty of integrating BAW and SAW sensors into an electronic nose resides in the more complex electronics and their high sensitivity to disturbances such as temperature and humidity fluctuations [14].

6.4.5. Biosensors

A biosensor consists of an immobilised biologic molecule (enzymes, cellules or antibodies) next to a transducer, which transforms chemical signal into an electric signal or into other kind of output as optical, acoustic and heat signal when an analyte reaches to it. The biological element can be made of catalytic (enzymes, microorganisms and tissues) or non-catalytic components (antibodies, receptors and nucleic acids).

The transduction element allows the transformation of the chemical signal, obtained from a biological process, into another kind of signal. According to the kind of the transduced signal, biosensors can also be grouped in electrochemical (amperometric, potentiometric, conductimetric), optical, calorimetric and acoustic biosensors. The biosensor sensing mechanisms depend on the transduction technology. In the case of electrochemical biosensors for example, the transduction system is based on an electrochemical nature process, and so the transduced signal is electric. Most of the commercial biosensors are electrochemical; because the electrochemical instrumentation is relatively simple and inexpensive. The current flows at constant potential with respect to a reference electrode in the amperometric sensors, and the current generated by the oxidation or reduction of electroactive species at the surface of the working electrode is measured [13].

6.5. Application Areas of E-nose

Recently, e-nose has been developed for various applications, such as indoor air-quality monitoring, medical care, customs security, food quality control, environmental quality monitoring, military applications, and hazardous gas detection as shown in Fig. 6.4. [6].

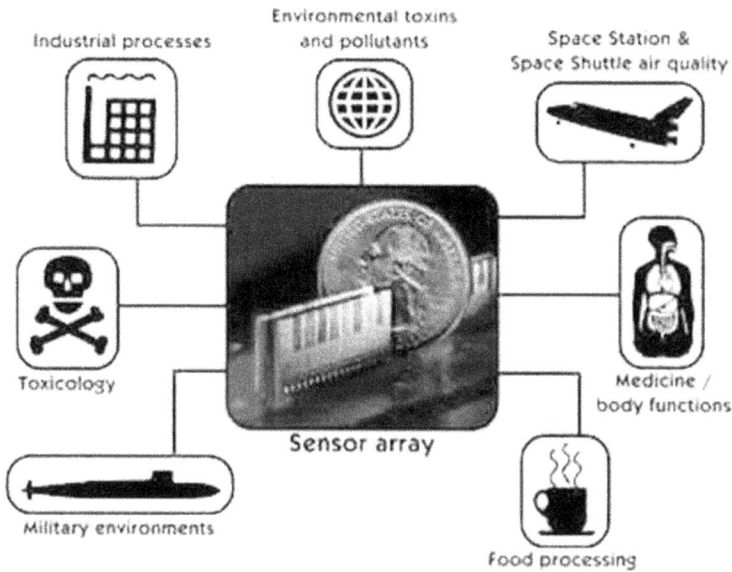

Fig. 6.4. Application areas of e-nose [15].

6.5.1. Medical Diagnosis and Health Monitoring

Medical diagnosis and health monitoring by e-nose include the following items:

6.5.1.1. Respiratory Disease Diagnosis

Human breath contains thousands of volatile organic compounds (VOCs) in gas phase.

E-nose can diagnose respiratory infections such as pneumonia. It does so by comparing smell prints from the breath of a sick patient with those of patients with standardized readings. It is also being studied as a diagnostic tool for lung cancer.

6.5.1.2. Cancer Detection

E-nose is capable of distinguishing the difference between the breath of a healthy person and a person with cancer. The device is especially promising because it will able to detect cancer before tumors become visible in X-rays.

6.5.2. Environmental Applications

- Analysis of fuel mixtures;
- Detection of oil leaks;
- Testing ground water for odours;
- Identification of household odours;
- Identification of toxic wastes;
- Air quality monitoring;
- Monitoring factory emissions etc.

6.5.3. Food Industry

- Monitoring of the food ripening (e.g., wine, cheese): Fruit ripening is associated with an accumulation of aromatic volatiles during ripening.
- Monitoring the storage of foods and food products (e.g., freshness and ageing control). Information from the noses can help in removal of rotten fruits at the appropriate time. This can help in avoiding storage losses due to rots and fruit diseases.
- Quality assurance of the selected raw foods
- Monitoring of the cooking processes
- Monitoring of the fermentation processes
- Monitoring other industrial processes (e.g., flavouring, blending, colouring).
- Quality assurance of the manufactured food products
- Monitoring of the product-packaging interactions
- Monitoring of the overall quality of the final food or food product

6.5.4. Crime Prevention

- E-nose is being developed for military and security applications in the detection of explosives and hazardous chemicals.
- To detect drug odours despite other airborne odours capable of confusing police dogs.

6.5.5. Space Applications (E-nose and NASA)

- It is a full-time, continuously operating event monitor used in the International Space Station.
- Designed to detect air contamination from spills and leaks in the crew habitat.
- Provides rapid, early identification and quantification of atmospheric changes caused by chemical species to which it has been trained.

6.6. Conclusions

An "electronic nose" is a system which is created to mimic the function of an animal nose. It offers a cheap and non destructive instrument that can be operated by non specialists. Since the whole process is automatic, the cost of each measurement is very low. The measurement cycle should be faster in order to increase throughput.

In fact, this analytical instrument is a "multi-sensor array technology" rather than a real "nose". It is still so far from the sensitivity and selectivity of a biological nose. Its aim can not be to replace the human nose totally.

References

[1]. K. Persaud, G. Dodd, Analysis of discrimination mechanisms in the mammalian olfactory system using a model nose, *Nature*, 299, 1982, pp. 352–355.
[2]. R. W. Moncrieff, An instrument for measuring and classifying odours, *J. Appl. Physiol.*, 16, 1961, pp. 742–749.
[3]. W. F. Wilkens, J. D. Hartman, An electronic analog for the olfactory processes, *J. Food Sci.*, 29, 1964, pp. 372–378.

[4]. T. M. Buck, F. G. Allen, M. Dalton, Detection of chemical species by surface effects on metals and semiconductors, in Surface Effects in Detection, *Spartan Books Inc.*, Washington, DC, USA, 1965.

[5]. J. W. Gardner, P. N. Bartlett, G. H. Dodd, H. V. Shurmer, Pattern recognition in the Warwick electronic nose, in *Proceedings of the 8th International Congress of European Chemoreception Research Organisation*, University of Warwick, UK, 18–22 July 1988.

[6]. S. W. Chiu, K. T. Tang, Towards a chemiresistive sensor-integrated electronic nose: A review, *Sensors,* 2013, 13, pp. 14214-14247.

[7]. J. W. Gardner, P. N. Bartlett, A Brief History of Electronic Noses, *Sensors and Actuators B,* 18, 1993, pp. 211–220.

[8]. E. Schaller, J. O. Bosset, F. Escher, Electronic Noses and Their Application to Food, *Lebensm.-Wiss. u.-Technol.,* 31, 1998, pp. 305–316.

[9]. E. Greenberger, Olfaction, Ch. 14 (https://www.studyblue.com)

[10]. J. W. Gardner, P. N. Bartlett, Electronic Nose. Principles and Applications; *Oxford University Press,* Oxford, UK, 1999.

[11]. M. Sahu, Seminar on Electronic Nose (https://www.slideshare.net)

[12]. A. Berna, Metal Oxide Sensors for Electronic Noses and Their Application to Food Analysis, *Sensors,* 2010, 10, pp. 3882-3910.

[13]. D. L. García-González, R. Aparicio, Sensors: From Biosensors to the Electronic Nose, *Grasas y Aceites,* 96, Vol. 53. Fasc. 1, 2002, pp. 96-114.

[14]. E. Schaller, J. O. Bosset, F. Escher, Electronic Noses and Their Application to Food, *Lebensm.-Wiss. u.-Technol.*, 31, 1998, pp. 305–316.

[15]. E-nose web portal (http://www.saba.kntu.ac.ir)

Chapter 7

Electronic Noses and Electronic Tongues

Tomasz Dymerski

7.1. Introduction

The human senses such as smell and taste are still main tools utilized in many areas of daily life, even the fact there were numerous attempts to elaborate a devices capable to perform their function. Nevertheless, the past 30 years of the evolution of sensor technologies gives a great opportunity to apply a newest technical solutions into the systems where sensor arrays are utilized. The electronic noses and electronic tongues are still being developed in order to measure aroma and taste in a way parallel to the biological senses. Despite of the fact these devices are not used for routine analysis, their prospective application might constitute a reasonable option too many issues where rapid analysis is advisable.

This contribution summarizes the achievements on the field of artificial senses, such as electronic nose and electronic tongue. It examines multivariate data processing methods and demonstrates a promising potential for rapid routine analysis. Main attention is focused on detailed description of sensor used, construction and principle of operation of these systems. A brief review about the progress in the field of artificial senses and future trends in concerned. A special attention has been paid to the application of these systems in two dominant fields, namely in food investigations and environmental monitoring.

Tomasz Dymerski
Department of Analytical Chemistry, Faculty of Chemistry, Gdańsk University
of Technology, Gdańsk, Poland

7.2. Senses of Smell and Taste

Gustatory sensations perceived by mammals originate when particles of volatile substances come into contact with taste receptors – specialized chemoreceptors clustered in taste buds that are located in the oral cavity. Clusters of these taste buds are located on small papillae which, depending on their location differ in shape and size. Adult humans have app. 10000 taste buds. Within each taste buds there are about 50-150 rod-shaped taste cells, which transmit the information to neuron cells which, in turn, transmits it to the brain. Five types of taste receptors which react to particular groups of chemicals present in foodstuffs or atmospheric air. Different taste sensations have different taste thresholds, highest for sweet and salty, and lowest for bitter foods. Taste sensations can be classified according to the taste-distinguishing mechanism into two groups. In the case of sour and salty tastes, the mechanism is based on hydrogen and sodium ions, respectively, reacting directly with ionic channels by changing the membrane potential of receptors [18, 23]. In the case of sweet and bitter tastes, there are protein receptor spots connected with the G protein, which, after forming a complex with a taste substance molecule, activate the G protein, leading to a series of chemical changes [4]. Both mechanisms lead to the excitation of a nerve pulse transmitted to the brain.

Sense of smell, together with the sense of taste, can be described as a 'chemical sense'. It is the ability to discern chemical compounds or their mixtures in the surrounding air [18, 24]. The working of the sense of smell can be divided into the following steps: learning to identify and differentiate flavours, remembering and integrating different odours [25]. The ability to integrate different flavours that is exhibited by the sense of smell is a very sophisticated function which allows for the prediction and correlation between components that will combine to create a new flavour [26]. The human olfactory apparatus is able to discriminate between app. 10 000 flavour compounds with high sensitivity and accuracy [15, 22], with sensitivity in the order of parts per trillion in the case of certain chemical compounds. In some cases, even stereoisomers can be distinguished [22]. For example methyl mercaptan, a compound responsible for the characteristic aroma of garlic, has an odour threshold of less than 500 pg/L of air [21]. However, for substances such as ethane, butane or acetylene the odour detection threshold is much higher [14]. Contrary to the taste of smell, it is difficult to classify aromas into distinct groups. It has been suggested that the number of basic classes of olfactory sensations varies from 7 to 50 [22].

Molecules of odorous substances are usually volatile (molecular masses up to app. 300 Da), small, polar and often hydrophobic [13, 14] organic compounds that contain one or two functional groups [27]. Simple odours, like that of ethyl alcohol, contain only one type of odorant molecule, whilst more complex ones can consist of several thousand different chemical components, each at a different concertation level [14, 15].

The sense of smell is one of the most primeval senses from the evolutionary point of view. The part of brain in which it is located is not dissimilar to its reptilian equivalent, which suggests that it developed before mammals have emerged as a distinct class of vertebrae. Mammalian olfactory receptors (OR) are seven-transmembrane domain G protein-coupled receptors and are encoded by a large number of genes. In the case of humans, 399 intact OR genes and 297 OR pseudogenes have been identified, which amounts to app. 3 % of the entire human genome [30, 31]. The human nose is app. triangular in shape, partitioned by an internal wall called the nasal septum. It is limited form the top by an orbital lamina perpendicular to the ethmoid bone, and from the bottom by a tetragonal copula [21, 26, 34]. The nasal cavity consists of two nasal tubes separated by a partition. The olfactory epithelium is located in the upper part of the nasal fossa, in the region of the nasal septum, the roof of the nasal cavity and at the front end of the superior nasal concha [21, 24]. Olfactory cells are neurons with a dual function, serving both as chemical receptors and pulse-conducting cells [21, 26, 34]. Olfactory cells are exposed to the atmospheric air, which makes them the only nerve cells in the human organism that receive the stimulus directly from the outside world. Only about 2 % of an aromatic substance reaches the olfactory epithelium which is a mechanism that protects ORs [21, 34]. In order to stimulate a single cell, less than 10 molecules are needed. Hydrophobic molecules are dissolved in the mucosa, which increases their concentration level. After a flavour molecule becomes bound to the protein cilia of the first neuron, redundant molecules are removed through mucus efflux, enzymatic degradation in sustentacular cells, and permeation to the intercellular space and to the vascular system [4, 21, 26]. From there, information is transmitted to the brain, where the analysis of sensation features is performed, allowing for discrimination of flavours and estimation of their intensity [24, 36].

The senses of taste and smell are functionally linked [4, 36] and connected with functions of the digestive system. This is demonstrated by the fact that a temporary inhibition of the sense of smell, e.g. by extensive mucosa production during common cold infection leads to changes in perception of a meal's flavour. The sense of smell and taste was the basis for concept of the construction of devices which mimics human senses. A certain analogy in working principle is shown in Fig. 7.1.

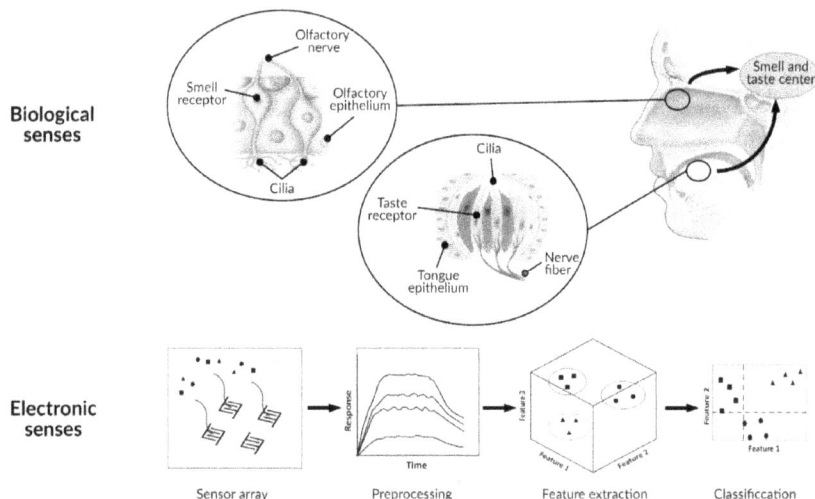

Fig. 7.1. Comparison of the sense of smell and taste with artificial senses.

7.3. Electronic Nose

An electronic nose is a device intended for the analysis of mixtures of odour substances, the operating principle of which is parallel to the mammalian olfactory sense. Devices of this type are also called odour-sensing systems [1], aroma sensors [2], electronic olfactometry [3], multisensory array technology [4], odour sensor, mechanical nose or artificial nose. At first, e-noses were equipped with arrays of sensors that could be considered analogues of olfactory receptors located in the human olfactory epithelium. Currently, devices based on different detection systems, e.g. fast gas chromatography or mass spectrometry are becoming increasingly popular [5]. For that reason, a more contemporary definition of electronic noses is an analytical device with the use of which it is possible to identify and classify mixtures of volatile chemical compounds in a relatively short time. The term 'electronic nose' was first proposed by Julian Gardner in 1988. However, the technique itself dates much further back. First devices of this type, albeit quite unsophisticated [6, 7], were introduced 40 years before the term was coined. In 1961 Moncreif created a device mimicking the olfactory sense equipped with 6 different sensors [8, 9]. First work describing a multi-sensory system equipped with chemical sensors was published by Persaud and Dodd in 1982, who were able to identify 20 odorants [10]. The golden age of the development of artificial senses came in 1990's, coinciding with major developments in the construction of chemical sensors. During this decade first commercial electronic noses were introduced to the market, e.g. Alpha M.O.S. in 1993, Neotronics and Aromascan in 1994 and Bloodhound and HKR Sensorsysteme in 1995. In 1998 mass spectrometry was used for the first time in this type of devices [11]. However, chemical sensors utilized in these early e-noses had some limitations, the most common being the need for frequent calibration, lack of stability, susceptibility to poisoning, response signal masking (e.g. by ethanol) and large power consumption. With the introduction of newer types of sensors, like MOSFET or piezoelectric sensors [12], some of these drawbacks were eliminated. When using MS-based electronic noses there is no issue with profile masking, sensor poisoning, and the impact of ambient conditions (e.g. relative humidity) on the response or non-linearity of the signal. Currently, the development of electronic noses is focused on increasing the sensitivity of the device's measuring array, which would allow to detect subtle differences between similar samples, and on miniaturization, limiting the power consumption and design of portable devices.

During the analysis using an electronic nose the output signal consists of responses of individual chemical sensors which, when considered holistically, create a characteristic aroma profile of the analyzed sample. Such aroma profile is unique for each mixture of aroma compounds and is sometimes called a 'fingerprint' or 'smellprint'. The identification of odour substance is achieved by comparing its fingerprint with the response to a reference sample [4]. There are many parallels between the mode of operation of an electronic nose and olfactory sense. Chemical sensors work like olfactory cells and send signals to the pattern recognition system which, similar to brain, identifies the aroma [13, 14]. E-noses ale usually comprised of an array of sensors placed in an insulated chamber, a sometimes thermostated sample chamber, electronic components (e.g. analogue-digital converter), and a pneumatic system (vacuum pump, valves, flow meters) (Fig. 7.2). For data processing software capable of statistical analysis is used.

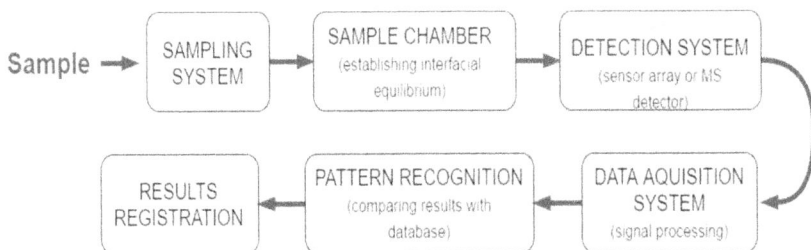

Fig. 7.2. Construction of electronic nose.

Sampling of volatile fraction can be done in a number of ways, depending on application and design of the device. Among others, direct headspace sampling, diffusion-based methods, bubbling or enrichment methods (e.g. solid-phase microextraction) are being used. A considerable challenge for constructors is the need to build each element which comes into contact with the sample out of an odourless, non-adsorptive, chemically inert materials. Both the sensor chamber and its volume should be designed in such a way, as to leave enough room to generate headspace [15]. Adequate sample incubation can be used to increase the concentration of volatile compounds in the headspace. Since the response of many types of sensors is influenced by changes in temperature and relative humidity it is necessary to monitor these parameters. Sample's headspace is carried by an inert gas or by ambient air to the sensor chamber [16]. The pump, characterized by low pulsations, is usually placed after the sensor chamber and operates in the vacuum (suction) mode, which reduces the risk of the analytes adsorbing on its elements. Before each analysis it is necessary to establish a base line of sensor's responses and measure the background noise. This is usually achieved by directing a stream of ambient air or inert gas through the sensor chamber [17, 18]. Chemical entities present in the gas come into contact with active elements of chemical sensors which leads to a change in their response signal due to, e.g. a change in conductivity. Intensity of the signal depends on the type of chemical substance, its affinity to the sensor's active element, and on its concentration. Collected response signals which form the distinct fingerprint are then processed using statistical methods. Chemometric techniques are used for data analysis (see sub-chapter 6). The comparison of the fingerprint with the odour profile of previously measured standards allows for identification and classification of the analyzed sample.

7.4. Electronic Tongue

The history of electronic tongue and nose starts in the beginning of the 20th century. Then the ion exchange theory was developed which has resulted in the construction of glass membrane electrodes used for measuring pH [19]. In the following twenty years, new sensors were designed, inter alia, MOSFET, BAW, ion exchange membrane, potentiometric biosensor, ISFET, PdMOS and SAW. In 1985, Otto and Thomas presented the first system for liquid phase analysis by using a multisensor array [20]. Seven years later, at the University of Kyusho the taste sensor was constructed by Toko; it consisted of ion-selective lipid membranes immobilized in PCV polymer [22, 23]. In 1995, as a result of cooperation between Russian and Italian research groups, a concept of electronic tongue was presented. It was based on inorganic chalgogenide glass sensor which enables both qualitative and quantitative analysis [23]. The most important facts related with development of both electronic tongues and electronic noses was listed in Table 7.1.

An electronic tongue is based on the same principles as electronic noses, with the most significant difference being that they are used to analyse solutions (mostly aqueous) instead of their volatile fractions. It is dedicated to identify, classify and analyze in a qualitative and quantitative way mixtures by applying a fingerprint method [6, 7]. An electronic tongue consists of three main elements, i.e. the sample-dispensing chamber or autosampler (optional), an array of sensors of different selectivity, and software for data processing (recognition system which mimics the brain functions) (Fig. 7.3) [30].

Application of this technical solution is consistent with the purpose of use of the human sense of smell. The array of sensors is mostly comprised of potentiometric, voltammetric, ion-selective field-effect transistor (ISFET), piezoelectric, and optical sensors. A major advantage of this type of devices is the fact, that they can be used to analyse liquid samples without any prior preparation, and solid samples after a simple dissolution [32, 33]. The sensors comprising the detection module of an electronic tongue are often not permanently fixed in an array. Because of that, the composition of the sensor matrix can be easily modified, making it flexible and adjustable to specific tasks. A unique feature of e-tongue devices is the possibility to correlate the results obtained using this technique with the results reported by a human sensory panel, which is particularly important for the food industry. They can be used to supplement the work of trained panelists especially in a situation when it is uncertain whether the analyzed substance is safe to consume.

Table 7.1. Chronological listing of most important achievements related to the development of electronic noses and tongues.

Year	Achievement	Author(s)	Artificial sense	Ref.
1906	Relationship between pH and the potential measured in glass membrane	Cremer	e-tongue, e-nose	[19]
1909	Development of glass electrodes	Haber, Klemensiewicz	e-tongue, e-nose	[19]
1936	Commercial production of glass electrodes	Beckman	e-tongue, e-nose	[19]
1937	Nikolsky equation	Nikolsky	e-tongue, e-nose	[19]
1937	Crystalline membrane	Nikolsky	e-tongue, e-nose	[19]
1954	First gas sensor	Hartman	e-nose	[24]
1959	First MOSFET sensor	Kahng and Atalla	e-nose	[24]
1960	Semiconductor gas sensors based on metal oxide	Taguchi	e-nose	[25]
1961	First mechanical instrument for sensing odors , i.e. an array of six sensors with different coatings	Moncrief	e-nose	[26]
1964	Artificial odor-recognition system	Wilkens and Hartman	e-nose	[27]
1964	BAW sensors	King	e-nose	[27]
1965	Research on the modulation of conductivity and contact potential for the use in odor monitoring	Buck, Dravnieks and Trotter	e-nose	[28]
1967	Ion-exchange membrane	Ross	e-tongue	[19]
1969	Potentiometric biosensor	Guilbault, Montalvo	e-tongue	[19]
1970	Ion-Sensitive Field Effect Transistor (ISFET)	Bergveld	e-tongue	[19]
1973	Invention of a PdMOS sensor	Lundström	e-nose	[27]
1979	First gas sensor based on a SAW oscillator	Wohltjen and Dessy	e-nose	[29]
1982	First electronic nose	Persaud and Dodd	e-nose	[27]
1985	Multisensor array for liquid phase analysis	Otto and Thomas	e-tongue	[21]
1988	Definition of an electronic nose	Gardner	e-nose	[27]
1992	Taste sensor	Toko	e-tongue	[22]
1995	Electronic tongue	Vlasov, Legin, D'Amigo, Di Natale	e-tongue	[23]

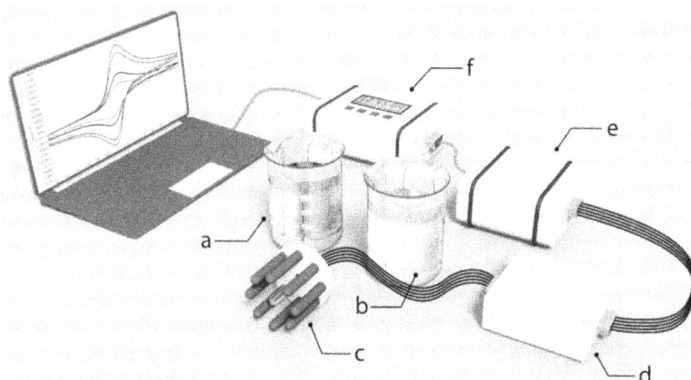

Fig. 7.3. Rendered view of electronic tongue setup: a – sample;
b - electrode rinsing solution; c - array of electrodes; d - relay box;
e – multiplexer; f – potentiostat.

Electronic tongues can be set up as either desktop or mobile versions. They can also be either static, i.e. when the sensors are submerged in static liquid, or dynamic, with the liquid pumped through the sensor chamber. There is usually also a second chamber used to rinse the sensor array after each analysis. The liquid containing analytes is often thermostated in order to maintain stability of measurement and repeatability. The types of chemical sensors used in electronic tongues do not differ in a significant way from the previously discussed sensors employed in electronic noses. One distinct category of devices are the so-called 'bioelectronic tongues', in which enzymes selective for particular compounds are utilized [33]. Arrays of biosensors are characterized by high selectivity due to enzyme-substrate interactions. Their efficiency can be further improved by facilitating the transfer of electrons from enzymes to the electrode (Downey). For that purpose electron mediators are introduced to the liquid. Devices equipped with arrays of biosensors have been used to study, among others, alcoholic beverages [34], fruits [35] and aging process in beer [36]. Besides biosensors, potentiometric and impedimetric sensors are also used in electronic tongues. The former are based on the voltage measurement at null current, which is needed in order to retain the balance of electrochemical processes [37]. They are relatively inexpensive and easy to manufacture, and also can be selective. On the other hand, fluctuations of temperature have a significant impact on their response signal. They are also prone to changes in potential caused by adsorption of components of solution on their active surface [6, 27, 39].

7.5. Sensors Used in Electronic Noses and Electronic Tongues

During measurement process an initial form of energy (an input signal) is transformed into a different form of energy, such as: electric, magnetic, chemical, thermal or radiation energy, which constitutes an output signal [39]. The element using which the measured parameter is read is called 'senor'. Together with certain type of signal converter enables the process of described transformation [28, 40]. Sensor is the most important and primary element in electronic noses and electronic tongues. There are two main critical properties of sensors, which makes them more or less useful in dedicated application. It is sensitivity and selectivity to odorants or taste substances that may be present in the analyzed samples. Sensors can be divided into four groups, i.e. electrochemical (conductometric, potentiometric, amperometric, voltammetric, and impedimetric), piezoelectric, optical sensors and biosensors. In case of electronic noses, instead of array of chemical sensors the proper detection system can be also applied. Some of this systems are equipped with gas chromatograph and/or universal detector, e.g. flame ionization detector, ion mobility spectrometer or mass spectrometer. Sometimes as a substitute of the detector the SAW sensor can be applied [27].

7.5.1. Conductometric Sensors

Conductometric sensors is not very often used in electronic tongues [33]. Their use is primarily in electronic noses. Their principle of operation is based on the changes in conductivity. These changes result from the interactions with the volatile odorants which leads to the changes in the sensor's electrical resistance. Despite the fact there are various types this sensors, the construction and distribution of specific elements in conductometric sensors is, in principle, the same. Conductometric sensors can be divided to types, i.e., Metal Oxide Semiconductor (MOS), Metal-Oxide Semiconductor field-Effect Transistor (MOSFET) and Conductive Polymer (CP) sensor [18, 28].

By far the most commonly used type of sensors are metal oxide semiconductors (MOS). They are popular because of their durability, relatively low price, sensitivity in the range from 5 to 500 ppm, chemical resistance and ease of use [42, 43]. Their active components are usually made of zinc, tin, tungsten or iridium oxides doped with platinum or palladium. When an analyte comes into contact with the active surface the sensor's resistance changes due to a catalytic reaction. The operating temperature of MOS sensors is in the range of 300 °C-500 °C which shortens the response and recovery time and prevents water vapour from condensing on their surface [15]. Unfortunately, this results in high power consumption, which is why this type of sensors is rarely used in portable devices. Moreover, they are susceptible to poisoning with sulphur compounds which create stable bonds with metal oxides [7, 26].

Metal oxide semiconductor field-effect transistors (MOSFET) are similar to MOS sensors, but operate at lower temperatures (100 °C - 200 °C). They are manufactured by depositing a catalytically sensitive metal in a gaseous phase on a silicon oxide surface. When analytes come into contact with the catalyst a reaction takes place, during which the sensor's physicochemical properties are modified, which in turn leads to a change in electrical potential. The sensors response signal magnitude is proportional to the concentration of analytes in the gaseous mixture. It is possible to manufacture very small MOSFET sensors without sacrificing the repeatability of analysis [44]. Due to the fact that the number of MOSFET varieties available on the market is relatively small they are seldom used in commercial devices.

Sensors based on conductive polymers (CP) belong to the so-called cold conductometric sensors. They are made of semiconducting materials set between two gold-plated electrodes. After the interaction with volatile molecules, reversible changes in conductivity are observed in the sensor. CP sensors are mainly sensitive to polar compounds. However their selectivity and sensitivity can be modified by introducing different functional groups to the polymer structure [17, 45]. Due to such modification, polymeric or glass fiber composites may also display high sensitivity to non-polar substances [46, 47]. The advantages of CP sensors are a low price and fast response, while susceptibility to humidity is the main disadvantage. These sensors have been used in an electronic nose to, inter alia, identify the stages of wine fermentation, monitor decomposition in the Atlantic salmon during its storage at different temperatures and detect the spoiled vacuum-packed beef. Cyranose 320, Aroma Scan A32/50S and Bloodhound BH114 are commercially available electronic noses that employ CP sensors [48].

7.5.2. Potentiometric Sensors

Potentiometric sensors are main elements of electronic tongues. These devices are used to monitor cheese fermentation [49], evaluate the impact of micro-oxygenation and oak chip maceration on the wine composition, in particular on the presence of phenol compounds [50], monitor changes during beer brewing [51], and identify the botanical origin of honey [52]. This principle of operation is based on the voltage measurement at null current, which is usually needed to retain the balance of electrochemical process. The generated signals are the electromotive forces which depend on the analyte activity. These signals originate in the process of electrochemical reaction at the electrodes or in solution. Potentiometric sensors have a lot of advantages, such as: low cost, commercially easy to produce, possibility to obtain selective sensors, and the highest degree of similarity with the mechanism of molecular recognition. Main disadvantages of this sensors are resulted by the dependence of the measured value on temperature and adsorption of the solution components onto the electrodes [52].

7.5.3. Amperometric and Voltammetric Sensors

This type of sensors is often used in electronic noses. Their use enables to assign the correct class to wheat samples in accordance with the quality classification [53]. They are also applied in electronic tongue systems to assess the conditions under which olive oil is stored [54], identify white wines with regard to the type of grapes and geographical origin [55], and discriminate among various blends of fruit juices [56]. Additionally, similarly to potentiometric sensors, these sensors are used to monitor the aging phase of wine [57], monitor beer fermentation [57], control the freshness of milk stored at room temperature [58], detect chemically adulterated red wine [59], and identify rice wine with regard to its age [60]. The measurements by means of these sensors are based on the electric current reading between the working and reference electrode in an electrochemical cell, as a function of analyte concentration. When an analyte reacts electrochemically, i.e. it undergoes oxidation or reduction, the electric current is generated at the working electrode. The reaction usually takes place at a constant potential as controlled with a potentiostat [28, 40]. In general, the working electrodes are made of gold, iridium, palladium, platinum and rhodium [61]. The main disadvantage of such sensors is the lack of selectivity.

7.5.4. Impedimetric Sensors

Impedimetric sensors are usually employed in an electronic tongues, however it does not happen often. This type of sensor was used to discriminate brands of red wines [60]. Their principles of operation is based on measuring the impedance at one constant frequency or for a frequency spectrum by means of impedance spectroscopy [63, 64].

7.5.5. Piezoelectric Sensors

The working principle of piezoelectric sensors is based on a piezoelectric phenomenon. This group can be divided into surface acoustic wave (SAW) sensors and bulk acoustic wave (BAW) sensors which also include quartz microbalance sensors (QCM). The sensors mass changes when analytes adsorb on its surface, which in turn leads to change of frequency at which the measuring element resonates [14]. This type of sensors is commonly used for measuring temperature, pressure, force or acceleration [17, 65]. The main difference between BAW and SAW sensors is, that in the former the generated wave travels through its entire volume, while in the latter only along its surface [17]. The main advantages of piezoelectric sensors are high sensitivity, real-time measurements, small size, durability, low cost. Piezoelectric sensors are commonly used in electronic noses to determine the optimal time for harvesting apples [65] and evaluate the quality of tomatoes [66]. Their wider use in electronic tongues is just a question of time.

Bulk acoustic wave sensors are comprised of a single quartz crystal and two gold-plated electrodes [15]. The crystal is coated with a selective layer which absorbs measured chemical substances from the gaseous mixture. When this happens, the resonator's weight increases and its resonating frequency, which ordinarily oscillates between 5 to 30 MHz changes. This gives rise to a three-dimensional acoustic wave which passes along the entire bulk of the crystal. Purging the crystal with an inert gas causes desorption of analytes and a return to default frequency [67]. The use of QCM sensors has numerous advantages, namely high sensitivity (ppb level), linearity of response at a large range of concentrations and insusceptibility to fluctuations of temperature and relative humidity.

Surface acoustic wave sensors have a higher resonating frequency of app. 100 MHz – 1 GHz. In this type of sensors a piezoelectric material with input and output transducers situated at its surface is used. A selective membrane is placed between the transducers. The membrane is usually made of porous polymers, lipids, self-assembled monolayers or Langmuir-Blodgett films. A two-dimensional acoustic wave travelling along the sensor's surface is generated when alternating current is directed through the input. The piezoelectric element is made of zinc oxide, lithium niobate or quartz. SAW sensors are less sensitive than QCM, since they react to concentrations at ppm level, but their response time is relatively short [15].

7.5.6. Optical Sensors

The use of optical sensors makes possible to detect substances which are not electrochemically active and therefore cannot be detected by an electrochemical sensor. Optical sensors have some disadvantages which significantly limit their application, for example, sensor durability and the output signal distortion [67, 69]. These sensors have been employed in electronic noses to discriminate commercial drinks [68], while in electronic tongues to evaluate the quality of beer brands [69] and discriminate wines with regard to the wine age and grape variety [59]. Optical sensors are devices mostly based on the interaction of electromagnetic radiation with matter. These sensors contain various dyes which react with analytes. The sensor sensitivity depends on the type of dyes used, a plethora of dyes can be employed in electronic senses [46, 70]. Optical sensors can be based on different phenomena such as, fluorescence, reflection and absorbance. These sensors consist of an indicator, detector and the light source, the latter set at a specific wavelength in order to maximize selectivity [17, 18]. The sensors are placed on special, mostly polymeric-type membranes. Due to the interaction with an analyte, the properties of an indicator change, which influences the membrane absorbance or fluorescence. The changes are monitored via a detector, which converts the signal from optical into electric form. There is a big variety of optical sensors therefore they are characterized by low cost, simple procedure and high selectivity.

7.5.7. Biosensors

Biosensors are mainly used in electronic tongues. These sensors have been used to monitor changes occurring during the aging process in alcohols [71]. A biosensor consists of a biological measuring element, which is located close to the transducer in order to achieve high sensitivity to the target analytes (Fig. 7.4).

Biosensor-based electronic tongue is often called a bioelectronic tongue. Until now, different principles of operation of bioelectonic tongues have been proposed, including voltammetric, amperometric and

166

potentiometric principle. One of the first voltammetric biosensor comprised three biocomposite electrodes that contained glucose oxidase and various metals such as, platinum, palladium and silver [71]. The next developed biosensor was based on the combination of potentiometric sensor and the enzyme urease, which had been covalently bound to carboxylated poly(vinyl chloride) [42, 63]. Another proposed solution was to combine the amperometric sensor, based on carbon and platinum working electrode and Ag/AgCl reference electrode, and the enzymes cholinesterases, tyrosinases, peroxidases and cellobiose dehydrogenases [62]. Biosensor arrays display high selectivity due to enzyme-substrate interactions. Moreover, the biosensor efficiency can be improved by the introduction of electron mediators, which facilitate the transfer of electrons from the enzyme to the electrode [72].

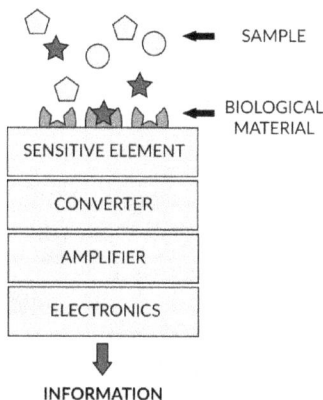

Fig. 7.4. Construction of biosensor.

7.6. Data Analysis Methods

The dataset obtained with the use of electronic nose or electronic tongue contains the response signals of each sensor and usually is very complex. Analysis of this type of data is considerably more difficult, than in the case of a device equipped with only one sensor. For that reason the first step of data processing is usually meant to decrease the dimensionality of the dataset. Doing this whilst retaining as much significant information as possible is one of key challenges in statistical data processing, as the results of data analysis should lead to reliable and repeatable results. Chemometric methods used for data processing utilize pattern recognition. Information contained in the sensor's response

167

signal is compared with reference data. The basic steps of data analysis are as follows [17, 74]:

- Pre-processing;
- Selection of variables;
- Classification;
- Decision making.

Preliminary analysis (pre-processing) is used to smoothen the signal, average sensor responses, and to filter the background noise [74]. Moreover, at this stage operations like data normalization and centering are also performed. Next step is to reduce the dataset's dimensionality and to employ one or more data analysis techniques. These techniques can be classified into quantitative methods and pattern analysis [76, 77] or, based on the model's learning mechanism into supervised and unsupervised methods. In supervised techniques, in order to properly 'teach' the model, it is fed input values together with expected outputs. In this way a reference signal is defined. In unsupervised methods the model is built without pre-defined outputs.

One of the easiest data analysis methods is to plot raw data as histograms or radar plots. This method works well when there are significant differences between groups of samples [74, 78, 79]. When it is difficult to classify samples outright, methods like multifactorial analysis or neurocomputers are employed [80–82]. In this chapter, the briefly characterization of the most commonly used statistical data analysis methods in combination with electronic noses and electronic tongues are presented.

7.6.1. Artificial Neural Networks (ANN)

Models utilizing artificial neural networks can work both as supervised and unsupervised pattern recognition techniques. Artificial networks are comprised of interconnected computation clusters, so-called 'nods'. In each node a simple mathematical operation that modifies the input is performed. This solution mimics the processes taking place in human neural cells. Nods are usually situated in layers. The task of the first strata is to relay the information to other nodes. Subsequent nodes modify the signal and their output is the input of other clusters. In this way, neural network 'learns' how to react to a given impulse, e.g. a mixture of odour compounds [82]. At this stage weights are determined, that is by what factor the signal will be multiplied in a given node. Nodes situated in the last layer ale called output neurons. A model set up in this way can be used to predict output values, classification or for pattern recognition [83]. The most commonly used Chemometric methods based on ANN are radial basis function (RBF), counter propagation – artificial neural network (CP – ANN), generalized regression neural network (GRNN), time delay neural networks (TDNN), probabilistic neural networks (PNN) and self-organizing maps (SOM) [84].

7.6.2. Cluster Analysis (CA)

In this method inputs are classified based on the distances between individual data points [85]. The result of cluster analysis is often plotted in the form of dendrograms [86]. Each strand is assigned to gradually larger agglomerates based on the similarity of data (distance between particular strands). In the end each object forms a single cluster [85]. Clusters can be grouped not only based on distance to the nearest neighbour, but also based on the distance from the furthermost neighbour or group average [86]. Each object in the dataset can be assigned to only one class. In practice, CA can be used to determine, whether or not there are distinct sub-classes in a given group [87].

7.6.3. k-Nearest Neighbour (k-NN)

K-nearest neighbour algorithm classifies data points based on distances between them in a multidimensional space. The points that are closest to each other create a separate group. Input data is compared with regard to Euclidean distances and therefore k objects that are most alike are selected. The 'k' parameter is chosen in such a way, as to maximize class separation [88].

7.6.4. Discriminant Function Analysis (DFA)

In this method discriminant functions are defined in such a way as to maximize the variance between various classes, and at the same time minimize the distances within the classes themselves [89]. This is achieved by using discriminant prediction equation, which allows to discard the variables that have the least impact on classification. In practice, first the F-test is performed in order to determine whether the discriminant model is entirely significant, and subsequently when the F-test has shown significance [90].

7.6.5. Partial Least Squares (PLS)

Partial least squares method is a modification of principal component analysis. It is a good example of a multiple regression method. The algorithm tries to fit a linear combination of predictors. This predictors, just like principal components, are orthogonal to each other. New variables have to explain the input data and be correlated with output values. PLS is employed when a single response signal can be explained by a large number of variables. It can be used to develop discriminant analysis models (PLS-DA) [91].

7.6.6. Linear Discriminant Analysis (LDA)

A probabilistic classification technique, LDA is based on rotating an orthogonal data space in such a way, as to maximize the variance between classes, and minimize the variance within those classes [82], [87]. Classes should have a normal distribution, there should be a linear separation of groups and variance-covariance matrices, and the number of objects should be at least three times larger than the number of variables [92, 93]. At the model training stage outputs of classes are established, add then input data is assigned to these classes [87].

7.6.7. Soft Independent Modelling of Class Analogies (SIMCA)

In SIMCA a separate model is created for each class. To this end e.g. principal components can be used [92, 94]. Then, based on a pre-defined similarity, a confidence envelope is created which allows to determine whether an object belongs to a particular group or not. The number of principal components used to create a model varies with each class.

7.6.8. Support Vector Machines (SVMs)

Support vector machines is an increasingly popular method of data processing in electronic noses. It was first proposed by Vapnik in 1992 [95], which makes it one of most recent supervised learning techniques. In it, points in a multi-dimensional space are separated by a hyperplane to form distinct classes. How successful this separation is depends on hyperplane's confidence margin, defined as the distance between he hyperplane and object closest to it (so-called support vectors) [96, 97]. Variants of this method are also used, such as support vector regression or least squares support vector regression (LSSVR) [98].

7.6.9. Analysis of Variance (ANOVA)

This method is used to compare more than two groups, classification of which is based on a single variable (one-way ANOVA). In this parametric tool the variance of dependent variables is compared within groups which were created based on independent variables. In order to employ the analysis of variance it is necessary to know the dependent variables that are characterized by the normal distribution and by being measureable at least on an interval scale. Moreover, it is important for the groups of variables to be comprised of the same number of objects [99].

7.7. Applications

To date, a great number of studies devoted to various applications of electronic noses and electronic tongues are reported. This artificial senses were used in many different areas of everyday life, e.g. in chemical industry including explosive materials, petrochemical, cosmetic, pharmaceutical, paper, packaging industries, as well as liquid-gas distribution and bootling plants, criminological and medical tests, agriculture, transportation, manufacturing and military purposes. In respect of many years of using this technology the general conclusions can be deduced. Firstly, rather more applications are devoted to electronic noses than electronic tongues and secondly, the main area of interest of artificial senses during many years and now is still food industry. Second main application is environmental monitoring, which was the first use of electronic sense, namely electronic nose to analyze the mixtures of volatile air contaminants that had already been detected by olfaction [72]. Therefore, in present subchapter this two above-mentioned and key applications will be discussed, despite of the fact that electronic tongue utilization in second application is minor.

7.7.1. Food Investigations

In the case of food process monitoring, the electronic nose and tongue have been used to monitor fermentation in milk and cheese [50, 101]. The application of artificial senses assures the product quality at the very start of the food production line. The monitoring of food freshness and product quality during storage is the foremost application of the electronic nose and electronic tongue [32, 59, 81, 102–109]. This allows the exclusion of spoiled products from the market as well as the determination of appropriate shelf times and storage conditions for milk and cheese in order to avoid financial losses by the dairy industry. Chemical analyses of meat products are mainly performed via an electronic nose, the use of electronic tongue in this case is less suitable because it requires a more complex preparation of samples [31], medical condition of meat [102, 103], and the influence of storage conditions on meat quality in order to determine expiration date [111]. The electronic tongue was applied to analyze ground meat in order to predict the level of chlorides, nitrates and nitrites [110]. Actually, it is the electronic nose which has found a broader application in the field of meat product analysis. This device is used to, among others, monitor the curing process in Iberian ham in order to detect spoilage Fish and shellfish [111]. The electronic senses are employed in the fish processing industry to mainly evaluate the freshness of fish and shellfish. The electronic nose has been used to perform such analysis in sardines [112, 113], shrimp and cod roe [59, 72], and the Atlantic salmon [114], while the electronic tongue, on samples of bream [115]. The duration and conditions of storage have big influence on the freshness of fish and shellfish. In order to protect consumers from the purchase of old fish, studies on the relationship between the duration and conditions of storage, and product freshness were conducted by means of an electronic nose on cod fillets [116], fresh and frozen Atlantic salmon, tilapia [116], the Argentine hake [117], and oysters [118]. An electronic nose is also used to analyze fruits and vegetables. The application of this device is related to the monitoring of food processing, and the evaluation of freshness, shelf life and authenticity of food. On the other hand, the electronic tongue has been mostly used to classify cultivars (e.g. discrimination between onion and shallot) [119], tomatoes based on various parameters [120], apples [121] and apricots, the latter being also discriminated based on the storage duration [121]. The following fruits and vegetables have been investigated so far: bananas, apples, pears, oranges, strawberries, broccoli, potatoes and carrots. The electronic nose was used to monitor dehydration in tomatoes [122] and grapes [123] in order to determine the optimal storage time. Olive oil is very frequently analyzed because the attempts to adulterate more expensive olive oil with its cheaper alternative are very common [123]. Therefore the electronic tongue and nose were both used to discriminate among olive oils based on oil geographic origin [124], and type and quality [55, 125]. In the case of olive oil, the conducted studies were also aimed at determining rancidity [126] and the relationship between the storage time and oil quality [54]. In the case of grains and grain products, quality evaluation and authentication are important (wheat [53, 29], rice [127], barley and oats [128, 129], and corn [130]) however the critical issue for the consumers is the evaluation of possible health risks of such products by means of electronic senses. The possible health risks are related to grain diseases, which often go unnoticed due to the fact that they are not visible to the naked eye. Mycotoxin contamination of grain, which also includes contamination with aflatoxins, and various diseases caused by fungi and bacteria are such risks. Coffees, teas and herbal infusions are mainly analyzed in order to distinguish among specific types, quality levels and brands. This is due to the fact that these products are highly variable. The products of low and high quality are frequently mixed together to lower the overall production costs, and then sold as top quality merchandise. Coffees were discriminated based on the quality [131], brand [119, 132, 133] and the ripening period [134], while teas were evaluated in relation to quality [135], brand [136], geographical origin [136], and the content of flavoring substances such as, coumarin [137], theaflavin [138]. Moreover, the electronic nose was used to monitor the fermentation process in black tea, and to determine the optimal time for producing tea with the best flavor [138]. Until now, the studies aimed at identifying brands and quality of beverages were conducted in e.g. cola type drinks [138], other commercial beverages [139], mineral water [140], and fruit juices and fruit juice-based drinks [141, 142]. On the other hand, alcoholic beverages are among the products which have been most frequently analyzed by means of electronic senses. Process monitoring in alcohol production requires fast analytical tools which can detect substandard products, discriminate among products, and authenticate products in real-time. Wine and beer is mostly subjected to such type of monitoring because both beverages undergo fermentation that results in a release of specific compounds influencing the taste and aroma of the final product. The artificial senses monitored the processes of fermentation [142], brewing [51] and aging [71] in beer. In the case of wine, these devices were employed to control the aging process, determine the influence of wooden barrels on aging and maceration [51, 58], and monitor grape fermentation [143]. Authentication of alcohols is not only aimed at detecting adulterated products with substandard characteristics, but also at identifying falsified products that can be potentially harmful to the consumer's health. Until now, the electronic nose was used to discriminate among vodkas [144], spirits [145], whiskeys [146], wines [146], tequilas [146], beers [146] and sorghum-based drinks [147]. The examples of electronic nose and electronic tongue applications in food analysis are shown in Table 7.2.

Table 7.2. Application of electronic noses and electronic tongues in food investigations.

Application	Sample	Object of investigation	Method of data analysis	Ref.
Food process monitoring; E-NOSE	Milk and cheeses	Monitoring of the smell intensity during fermentation	PCA	[148]
	Cencara tomatoes	Monitoring of the dehydration process	PCA	[122]
	Black tea	Determination of the optimal duration of fermentation	TDNN, SOM,	[149]
Evaluation of food freshness; E-NOSE	Cod fillets	Discrimination of samples based on different storage times	PLS	[116]
	Smoked Atlantic salmon (fresh and frozen)	Classification of spoilage in samples at different temperatures	PLS, PLSR	[114] [150]
	Meat	Freshness evaluation in meat in relation to storage time and storage conditions	PCA, LDA, BPNN, CDA	[150]
	Olive oil	Detection of rancidification	PCA	[151]
Testing the shelf life of food; E-NOSE	Pink Lady and Jonagold apples	Discrimination of varieties ripeness level, shelf life and storage conditions	PCA, ANN, PLS,	[152] [153] [154]
	Milk	Determining the influence of storage time on milk	PCA,	[105] [155]
	Meat	Determination of shelf life	LDA	[101]
Authentication of food; E-NOSE	Tequila, whiskey, vodka	Discrimination of four types of beverages	PCA, DFA	[146]
	Italian wines	Detection of the falsification	PCA, BP/ANN	[156]
	Green tea	Discrimination of the quality classes in Longjing tea	LDA, PCA, ANOVA, ANN	[157] [135]
		Identification of coumarin-enriched green tea	PCA, CA	[137]
		Discrimination of the green tea brands	PCA, ANN	[137]
Other applications; E-NOSE	Olive oil	Discrimination of quality classes based on the qualitative and quantitative information	SOM, SIMCA, PLS PCA	[137] [158] [159]
	Cabernet red wine	Monitoring of changes in wine aroma after bottle opening	PCA, SOM	[160]
	Oranges, apples and peaches	Post-harvest quality evaluation	PCA, PLS, ANN	[161] [162]
Food process monitoring; E-TONGUE	Bacterial cultures used in the cheese production	Monitoring of the fermentation process	PLS	[49]

Application	Sample	Object of investigation	Method of data analysis	Ref.
	Red wines	Monitoring of wine aging	PCA, SIMCA	[163]
		Evaluating the influence of micro-oxygenation and oak chip maceration on wine composition.	PCA, PLS, ANOVA	[50] [164]
	Beer	Variability monitoring of the brewing process	PCA, PLS	[51]
		Monitoring of changes during the aging process in beer	PCA, LDA, RBF, PNN, BP-NN	[51]
		Monitoring of the fermentation process	PLS	[69]
Evaluation of food quality and freshness; E-TONGUE	Fillets of farmed gilthead seabream Sparus aurata	Discrimination of storage time. Predicting the parameters of spoilage	PCA, ANN, PLS MLR	[165]
	Tench (Tinca tinca)	Discrimination of storage time	PCA	[166]
	Apricots	Discrimination based on post-harvest storage time and apricot varieties	CDA, PLS	[166]
	Non-alcoholic beverages	Quality evaluation of high-fructose corn syrup	SIMCA	[51]
	Rice	Quality evaluation of milling	PCA	[167]
	Milk	Monitoring the freshness of milk stored at room temperature	PCA, ANN, CA, PLSR, LS-SVM	[58] [168]
Stability testing of food; E-TONGUE	Extra virgin olive oil	Evaluation of different storage conditions	LDA	[54]
Authentication; E-TONGUE	Honey	Discrimination based on botanical origin	PCA, PLS, ANN, LDA	[52] [169]
		Discrimination based on botanical and geographical origin	PCA, CA, CCA, ANN	[170] [171]
	Milk	Discrimination of milk samples pasteurized in different ways	PCA	[171]
	Alcoholic beverages	Fast quality evaluation of alcoholic drinks and identification of brands	PLS, LDA, PCA	[172]
		Determination of ethanol content in alcohols from different sources	PCA, SIMCA, PLS, PCR	[173]
		Discrimination between the cheap and expensive whiskies	PCA	[174]
Other applications; E-TONGUE	Extra virgin olive oil	Discrimination of samples based on phenolic compounds content	PCA, PLS, PLS–DA	[175]
	Black tea	Determination of theaflavin level	PCA, ANN	[138]
	Ground meat	Prediction of the level of chlorides, nitrates and nitrites	PLS	[138]

7.7.2. Environmental Monitoring

Electronic senses, namely electronic nose and electronic tongue, can be successfully used for monitoring environmental pollution. There is a need to develop devices capable of evaluating the state of the environment rapidly or even in real time, without supervision and at a relatively low cost. Electronic noses and tongues show promise in this regard and find application in monitoring the quality of water and of atmospheric air [176, 177]. The analysis of atmospheric air can be performed in several ways. One can, for example, measure the concentration of several pre-defined substances or analyse the air holistically. Both these tasks can be performed using sensor systems. One of the first investigations in this area was performed in the first half of 1990' [178]. In it, a device equipped with CP sensors was used to analyse an aqueous solution of ethanol, diacetyl and dimethyl sulphide. Electronic nose was also used to measure the concentration of nitric oxide, methane and carbon monoxide at 500-2000 ppm concentrations [179]. It is important to note, that when using e-noses to determine particular substances there can occur interferences caused by the presence of other chemical compounds. A research has shown, that when determining hydrogen sulphide and nitrogen dioxide in a mixture containing carbon dioxide and water vapour the presence of humidity and CO_2 had a significant impact on the sensor's response signal, but it was possible to properly identify the components of the gaseous mixture using discriminant factor analysis [180]. It is possible to use the electronic nose to determine certain VOCs at a very low concentration level (ppb level) [181, 182], even below the threshold limit value (TLV). That was the case with benzene, methanol, ethanol, toluene and acetone determined below TLV using a device equipped with MOS sensors [183]. Sensor drift poses a significant problem, leading to high measurement uncertainties. In most recent applications electronic noses are being mounted on mobile robots [184], but because of insufficiently advanced models of gas distribution this technology is yet to find real-life applications. Another important application of electronic noses is odour classification and odour intensity evaluation. Currently, the golden standard in determination of odour nuisance is dynamic olfactometry [185]. Using this technique it is not possible to determine the nature of the odorant. Moreover, it is not sufficiently sensitive and cannot be used to constantly monitor odour nuisance [177]. A solution can be to use sensor systems. In one of the first investigations in this area two electronic noses: Odourmapper and Aromascan PLC were used for odour quantification [186]. When investigating odour intensity, the results obtained using e-noses are similar in nature to those obtained using dynamic olfactometry. In one study that compared the two techniques when applied to poultry livestock nuisance monitoring has shown, in the range from 250 to 4500 oue/m^3 a linear correlation with R factor of 0.89 [187]. Classification of odorous air is also an important issue. Often an unpleasant odour, even at low intensity, is less acceptable than a pleasant one at high intensity. In 2007 the use of electronic noses for classification of odours in the vicinity of a composting plant was evaluated [188]. The accuracy index of classification was 72 %. Electronic olfaction can also be used for evaluation of water quality, e.g. to determine the biological oxygen demand (BOD) [189, 190]. Electronic noses were also used to determine pesticides [191], sulphides and nitrates [192] or to detect the presence in water of microorganisms responsible for emission of odours [193]. Other applications of e-noses in environmental protection include the control of processes taking place during composting or sewage treatment and monitoring of the operation of devices used for purifying water and gasses from undesirable substances, including odours [177]. The examples of electronic nose applications in environmental monitoring are presented in Table 7.3.

Due to their limitations electronic tongues are not used as often as electronic noses in environment monitoring. The main area of their application in this area is detection of water pollution, both potable and processed at sewage treatment plants. Electronic tongues can be used e.g. to monitor semi-volatile compounds like herbicides [176] or determinate of the contents of inorganic anions and transition-metal cations in model ground, mine and sea water.

7.8. Summary

Over last 30 years the evolution of sensor technologies gives a great opportunity to apply a newest technical solutions into the systems where sensor arrays are utilized. Electronic noses and electronic tongues in the assumptions are devices dedicated to measure aroma and taste in a way parallel to the human biological senses. This gives a great opportunity to apply them where human being is not able to make proper measurements with required repeatability, objectivity and sensitivity, and in the places where toxic conditions may exists. Due to the fact that these devices are electronic ones, it gives a huge potential for applications in which *on-line* or data transmission systems is considered.

Table 7.3. Examples of the application of electronic noses in analysis of air pollution.

Type of sample	Analytes	Data analysis method	Ref.
Air samples from composting plants, printing houses, sewage treatment plants, recycling plants, settlers	Water vapor, flammable gases, toxic gases, solvents	PCA, DA	[194]
Internal air samples	CO, NO_2, VOC (benzene, toluene, m-xylene)	SOM	[195]
Samples of air from the sewage treatment plant	VOC	PCA	[196]
Air samples from sewage treatment plants	VOC	SOM	[197]
Indoor air samples from duck farms	H_2S, NO_2, SO_2	DFA	[198]
Internal air samples	Ammonia, VOCs	ANN	[199]
Internal air samples	CO, NO_2, ammonia, VOC (benzene, toluene, formaldehyde)	HSVM, SVM, FDA, MLP	[200]
Samples of indoor air from cars	CO, NO_2, VOC (benzene, toluene, formaldehyde)	BPNN	[201]
External air samples of from Alexandria	CO, CO_2	ANOVA	[177]

This fits into contemporary canvas of national and international smart specialization strategy, especially when the use of telemedicine is expected. Moreover, these systems have even more advantages of which the possibility of their miniaturization and execution of rapid analysis is the key one. Therefore, despite of the fact that there are already many applications of electronic noses and electronic tongues, even more can be invented in the known areas and new ones. For example, the biosensors are more commonly used in electronic tongues than in electronic noses. Continuous evolution of sensor technology might, in the near future, lead to balance of this proportions. Furthermore, electronic nose and electronic tongue technology is on the verge of renesaince. It is understandable that the fastest progress in sensor development was years ago, nevertheless this technology is before the greatest development. The situation is analogical to one of the 50's. That time only a best specialist had a contact with computers. Nowadays, computer is well known device and available at every step. Perhaps the appropriate adaptation of artificial senses to the needs of the average user (not R & D person) will launch a new chapter in the field of electronic nose and electronic tongue technology.

Acknowledgements

This work was financially supported by the Grant No. PBSII/B9/24/2013 obtained from the National Centre for Research and Development of Poland and Grant no. 2015/19/B/ST4/02722 received from the National Science Center of Poland.

References

[1]. J. W. Gardner, P. N. Bartlett, A brief history of electronic noses, *Sens. Actuators B*, Vol. 18, Issue 19, 1994, pp. 221–220.

[2]. P. Mielle, 'Electronic noses': Towards the objective instrumental characterization of food aroma, *Trends Food Sci. Technol.*, Vol. 7, Issue 12, 1996, pp. 432–438.

[3]. M. P. Martí, O. Busto, J. Guasch, R. Boqué, Electronic noses in the quality control of alcoholic beverages, *TrAC Trends Anal. Chem.*, Vol. 24, Issue 1, 2005, pp. 57–66.

[4]. V. P. Shiers, Electronic nose technology-evaluations and developments for the food industry, in *Food Ingredients Europe: Conference Proceeding*, 1995, pp. 198–200.

[5]. A. D. Wilson, M. Baietto, Applications and Advances in Electronic Nose Technologies, *Sensors*, Vol. 9, Issue 7, 2009, pp. 5099–5148.

[6]. T. C. Pearce, Computational parallels between the biological olfactory pathway and its analogue 'the electronic nose': Part II. Sensor-based machine olfaction., *BioSystems*, Vol. 41, Issue 2, 1997, pp. 69.

[7]. E. Schaller, J. O. Bosset, F. Escher, 'Electronic Noses' and Their Application to Food, *LWT - Food Sci. Technol.*, Vol. 31, Issue 4, 1998, pp. 305–316.

[8]. G. Sujatha, N. Dhivya, K. Ayyadurai, D. Thyagarajan, Advances In Electronic -Nose Technologies, *Int. J. Eng. Res. Appl.*, Vol. 2, 2012, pp. 76–84.

[9]. A. Trinchi, An instrument for measuring and classifying odors, *J. Appl. Physiol.*, Vol. 16, 1961, p. 742.

[10]. K. Persaud, G. Dodd, Analysis of discrimination mechanisms in the mammalian olfactory system using a model nose, *Nature*, Vol. 299, Issue 5881, 1982, pp. 352–355.

[11]. B. Dittmann, S. Nitz, G. Horner, A new chemical sensor on a mass spectrometric basis, *Adv. Food Sci.*, Vol. 20, 1998, p. 115.

[12]. D. James, S. M. Scott, Z. Ali, W. T. O'Hare, Chemical sensors for electronic nose systems, *Microchim. Acta*, Vol. 149, Issue 1–2, 2005, pp. 1–17.

[13]. T. Toko, T. Murata, T. Matsuno, Y. Kikkawa, K. Yamafuji, Taste Map of Beer by a Multichannel Taste Sensor, *Sensors Mater.*, Vol. 4, 1992, pp. 145–145.

[14]. R. M. Kumar, V. R. Bhethanabotla, Sensors for Chemical and Biological Applications, *CRC Press*, 2010.

[15]. F. Korel, M. Ö. Balaban, Electronic Nose Technology in Food Analysis, in Handbook of Food Analysis Instruments, S. Ötleş (Ed.), *CRC Press*, Boca Raton, FL, 2009, pp. 365–378.

[16]. H. T. Nagle, S. S. Schiffman, R. Gutierrez-Osuna, The How and Why of Electronic Noses, *IEEE Spectr.*, Vol. 35, Issue 9, 1998, p. 22.

[17]. K. Arshak, E. Moore, G. M. Lyons, J. Harris, S. Clifford, A review of gas sensors employed in electronic nose applications, *Sens. Rev.*, Vol. 24, Issue 2, 2004, pp. 181–197.

[18]. A. K. Deisingh, D. C. Stone, M. Thompson, Applications of electronic noses and tongues in food analysis, *Int. J. Food Sci. Technol.*, Vol. 39, Issue 6, 2004, pp. 587–604.

[19]. Y. Vlasov, A. Legin, Non-selective chemical sensors in analytical chemistry: from 'electronic nose' to 'electronic tongue,' *Fresenius J. Anal. Chem.*, Vol. 361, 1998, pp. 255–260.

[20]. M. Otto, J. D. R. Thomas, Model studies on multiple channel analysis of free magnesium, calcium, sodium, and potassium at physiological concentration levels with ion-selective electrodes, *Anal. Chem.*, Vol. 57, 1985, pp. 2647–2651.

[21]. T. Toko, T. Murata, T. Matsuno, Y. Kikkawa, Y. K., Taste map of beer by a multichannel taste sensor, *Sens. Mater.*, Vol. 4, 1992, pp. 145–151.

[22]. K. Hayashi, Electric characteristics of lipid-modified monolayer membranes for taste sensors, *Sens. Actuators B*, Vol. 23, 1995, pp. 55–61.

[23]. A. Legin, A. Rudnitskaya, Y. Vlasov, C. Di Natale, F. Davide, A. D'Amico, Tasting of beverages using an electronic tongue, *Sens. Actuators B*, Vol. 44, 1997, pp. 291-296.

[24]. K. Mauldin, C. E. Puntambekar, Solution-Processable α,ω-Distyryl Oligothiophene Semiconductors with Enhanced Environmental Stability, *Chem. Mater.*, Vol. 21, 2009, pp. 1927–1938.

[25]. A. Trinchi, Investigation of sol–gel prepared Ga–Zn oxide thin films for oxygen gas sensing, *Sens. Actuators B*, Vol. 108, 2003, pp. 263–270.

[26]. R. Moncrieff, W. An instrument for measuring and classifying odors, *J. Appl. Physiol.*, Vol. 16, 1961, pp. 742–749.

[27]. T. M. Dymerski, T. M. Chmiel, W. Wardencki, Invited review article: an odor-sensing system-powerful technique for foodstuff studies., *Rev. Sci. Instrum.*, Vol. 82, Issue 111101, 2011, pp. 1–32.

[28]. M. García, M. Aleixandre, J. Gutiérrez, M. C. Horrillo, Electronic nose for ham discrimination, *Sens. Actuators B*, Vol. 114, Issue 1, 2006, pp. 418–422.

[29]. I. Y. Zayas, C. R. Martin, J. L. Steele, A. Katsevich, Wheat classification using image analysis and crush-force parameters, *Trans. ASAE*, Vol. 39, 1996, pp. 2199–2204.

[30]. K. Persaud, G. Dodd, Analysis of discrimination mechanisms in the mammalian olfactory system using a model nose, *Nature*, Vol. 299, 1982, pp. 352-363.

[31]. F. Winquist, I. Lundström, P. Wide, The combination of an electronic tongue and an electronic nose, *Sens. Actuators B*, Vol. 58, Issue 1–3, 1999, pp. 512-521.

[32]. L. L. Leake, Electronic Noses and Tongues, *Food Technol.*, Vol. 6, 2006, pp. 96-101.

[33]. M. L. Homer, Novel materials and applications of electronic noses and tongues, *MRS Bull.*, Vol. 29, Issue 10, 2004, p. 697.

[34]. P. Ciosek, Z. Brzózka, W. Wróblewski, Electronic tongue for flow-through analysis of beverages, *Sens. Actuators B*, Vol. 118, Issue 1–2, 2006, pp. 454-459.

[35]. E. A. Baldwin, J. Bai, A. Plotto, S. Dea, Electronic Noses and Tongues: Applications for the Food and Pharmaceutical Industries, *Sensors*, Vol. 11, Issue 5, 2011, pp. 4744-4750.

[36]. S. Magdalena, P. Wiśniewska, T. Dymerski, J. Namieśnik, W. Wardencki, Food Analysis Using Artificial Senses, *J. Agric. Food Chem.*, Vol. 62, Issue 7, 2014, pp. 1423–1448.

[37]. J. E. Haugen, K. Rudi, S. Langsrud, S. Bredholt, Application of gas-sensor array technology for detection and monitoring of growth of spoilage bacteria in milk: A model study, *Anal. Chim. Acta*, Vol. 565, Issue 1, 2006, pp. 10–18.

[38]. J. R. Stetter, W. R. Penrose, The Electrochemical Nose, *Electrochemistry Encyclopedia*, 2001.

[39]. A. Wilson, M. Baietto, Applications and advances in electronic-nose technologies, *Sensors*, Vol. 9, Issue 7, 2009, pp. 5099-5148.

[40]. M. Peris, L. Escuder-Gilabert, A 21st century technique for food control: electronic noses., *Anal. Chim. Acta*, Vol. 638, Issue 1, 2009, pp. 1–15.

[41]. A. K. Deisingh, D. C. Stone, M. Thompson, Applications of electronic noses and tongues in food analysis, *Int. J. Food Sci. Technol.*, Vol. 39, Issue 6, 2004, pp. 587-594.

[42]. A. K. Deisingh, Application of Electronic Noses and Tongues, in Sensors for Chemical and Biological Applications, M. K. Ram and V. R. Bhethanabotla (Eds.), *Taylor & Francis Group*, Boca Raton, FL, 2010, pp. 173–194.

[43]. J. Kośmider, B. Mazur-Chrzanowska, B. Wyszyński, Odory, *PWN*, Warsaw, 2005.

181

[44]. T. C. Pearce, S. S. Schiffman, H. T. Nagle, J. W. Gardner, Handbook of Machine Olfaction: Electronic Nose Technology, *Wiley-VCH Verlag GmbH*, 2003.

[45]. Y. S. Kim, Fabrication of carbon black–polymer composite sensors using a position-selective and thickness-controlled electrospray method, *Sens. Actuators B,* Vol. 147, Issue 1, 2010, pp. 137–144.

[46]. P. Rapiejko, The sense of smell, *Alergoprofil*, Vol. 2, 2006, pp. 4–10.

[47]. M. Huotari, V. Lantto, Measurements of odours based on response analysis of insect olfactory receptor neurons, *Sens. Actuators B*, Vol. 127, 2007, pp. 284–287.

[48]. W.-X. Du, C.-M. Lin, T. Huang, J. Kim, M. Marshall, C.-I. Wei, Potential Application of the Electronic Nose for Quality Assessment of Salmon Fillets Under Various Storage Conditions, *J. Food Sci.*, Vol. 67, 2002, pp. 307–313.

[49]. K. Esbensen, Fermentation monitoring using multisensor systems: feasibility study of the electronic tongue, *Anal. Bioanal. Chem.*, Vol. 378, 2004, pp. 391–395.

[50]. A. Rudnitskaya, L. M. Schmidtke, I. Delgadillo, A. Legin, G. Scollaryd, Study of the influence of micro-oxygenation and oak chip maceration on wine composition using an electronic tongue and chemical analysis, *Anal. Chim. Acta*, Vol. 642, 2009, pp. 235–245.

[51]. T. Tan, V. Schmitt, S. Isz, Electronic tongue: A new dimension in sensory analysis, *Food Technol.*, Vol. 55, 2001, pp. 44–50.

[52]. L. A. Dias, A. M. Peres, M. Vilas-Boas, M. A. Rocha, L. Estevinho, A. A. S. C. Machado, An electronic tongue for honey classification, *Microchim. Acta*, Vol. 163, 2008, pp. 97–102.

[53]. J. R. Stetter, M. W. Findlay, K. M. J. Schroeder, C. Yue, W. R. Penrose, Quality classification of grain using a sensor array and pattern recognition, *Anal. Chim. Acta*, Vol. 284, 1993, pp. 1–11.

[54]. M. S. Cosio, D. Ballabio, S. Benedetti, C. Gigliotti, Evaluation of different storage conditions of extra virgin olive oils with an innovative recognition tool built by means of electronic nose and electronic tongue, *Food Chem.*, Vol. 101, Issue 2, 2007, pp. 485–491.

[55]. L. Pigani, Amperometric sensors based on poly(3,4-ethylenedioxy-thiophene)-modified electrodes: Discrimination of white wines, *Anal. Chim. Acta*, Vol. 614, 2008, pp. 213–222.

[56]. V. Martina, Development of an electronic tongue based on a PEDOT-modified voltammetric sensor, *Anal. Bional. Chem.*, Vol. 387, 2007, pp. 2011–2110.

[57]. V. Parra, A. A. Arrieta, M. Fernández-Escudero, J.A. Iniguez, J. A. de Saja, M. L. Rodríguez-Méndez, Characterization of wines through the biogenic amine contents using chromatographic techniques and chemometric data analysis, *Anal. Chim. Acta*, Vol. 563, 2006, pp. 229–237.

[58]. F. Winquist, C. Krantz-Rulcker, P. Wide, I. Lundström, Monitoring of freshness of milk by an electronic tongue on the basis of voltammetry, *Meas. Sci. Technol.*, Vol. 9, 1998, pp. 1937–1946.

[59]. V. Parra, A. A. Arrieta, M. L. Fernández-Escudero, J.A. Rodríguez-Méndez, J. A. De Saja, Electronic tongue based on chemically modified electrodes and voltammetry for the detection of adulterations in wines, *Sens. Actuators B*, Vol. 118, 2006, pp. 448–453.

[60]. Z. Wei, J. Wang, L. Ye, Classification and prediction of rice wines with different marked ages by using a voltammetric electronic tongue, *Biosens. Bioelectron.*, Vol. 26, 2011, pp. 4767–4773.

[61]. M. Scampicchio, D. Ballabio, A. Arecchi, S. M. Cosio, S. Mannino, Amperometric electronic tongue for food analysis, *Microchim. Acta*, Vol. 163, 2008, pp. 11–21.

[62]. L. Escuder-Gilabert, M. Peris, Review: highlights in recent applications of electronic tongues in food analysis, *Anal. Chim. Acta*, Vol. 665, 2010, pp. 15–25.

[63]. A. Riul, R. R. de Sousa, H. C. Malmegrim, A. dos Santos, D. S. Carvalho, F. J. Fonseca, O. N. Oliveira, L. H. C. Mattoso, Wine classification by taste sensors made from ultra-thin films and using neural networks, *Sens. Actuators B*, Vol. 98, 2004, pp. 77–82.

[64]. C. Caliendo, E. Verona, A. D'Amico, Surface Acoustic Wave (SAW) Gas Sensors, in Gas Sensors, *Springer Netherlands*, Dordrecht, 1992, pp. 281–306.

[65]. S. Saevels, J. Lammertyn, A. Z. Berna, E. A. Veraverbeke, C. Di Natale, B. M. Nicolaï, Electronic nose as a non-destructive tool to evaluate the optimal harvest date of apples, *Postharvest Biol. Technol.*, Vol. 98, 2003, pp. 77–82.

[66]. F. Sinesio, Use of electronic nose and trained sensory panel in the evaluation of tomato quality, *J. Sci. Food Agric.*, Vol. 80, 2000, pp. 63–71.

[67]. Keith J. Albert, Cross-Reactive Chemical Sensor Arrays, *Chem. Rev.,* 100, 7, 2000, pp. 2595–2626.

[68]. A. Berna, Metal oxide sensors for electronic noses and their application to food analysis, *Sensors* Vol. 10, 2010, pp. 3882–3910.

[69]. M. Kutyła-Olesiuk, A. Zaborowski, P. Prokaryn, P. Ciosek, Monitoring of beer fermentation based on hybrid electronic tongue, *Bioelectrochemistry*, Vol. 87, 2012, pp. 104–113.

[70]. M. A. Ryan, A. V. Shevade, H. Zhou, M. L. Homer, Polymer–Carbon Black Composite Sensors in an Electronic Nose for Air-Quality Monitoring, *MRS Bull.*, Vol. 29, Issue 10, 2004, pp. 714–719.

[71]. S. S. Ghasemi-Varnamkhasti, M. Rodríguez-Méndez, M.L. Mohtasebi, C. Apetrei, J. Lozano, H. Ahmadi, S. H. Razavi, J. A. de Saja, Monitoring the aging of beers using a bioelectronic tongue, *Food Control*, Vol. 25, 2012, pp. 216–224.

[72]. Q. Dong, L. Du, L. Zhuang, R. Li, Q. Liu, P. Wang, A novel bioelectronic nose based on brain-machine interface using implanted electrode

recording in vivo in olfactory bulb., *Biosens. Bioelectron.*, Vol. 49, 2013, pp. 263–269.

[73]. T. C. Pearce, S. S. Schiffman, H. T. Nagle, J. W. Gardner, Handbook of Machine Olfaction: Electronic Nose Technology, *Wiley-VCH*, 2003.

[74]. A. Ortega, S. Marco, T. Šundic, J. Samitier, New pattern recognition systems designed for electronic noses, *Sensors Actuators B Chem.*, Vol. 69, Issue 3, 2000, pp. 302–307.

[75]. H. T. Nagle, R. Gutierrez-Osuna, S. S. Schiffman, The how and why of electronic noses, *IEEE Spectr.*, Vol. 35, Issue 9, 1998, pp. 22–31.

[76]. F. Röck, N. Barsan, U. Weimar, Electronic Nose: Current Status and Future Trends, Vol. 4, Issue 3, 2007, pp. 34-43.

[77]. S. Scott, D. James, Z. Ali, Data analysis for electronic nose systems, *Microchim. Acta*, Vol. 2, Issue 1, 2006, pp. 112-119

[78]. W. Wardencki, T. Chmiel, T. Dymerski, Gas chromatography-olfactometry (GC-O), electronic noses (e-noses) and electronic tongues (e-tongues) for in vivo food flavour measurement, *Instrumental Assessment of Food Sensory Quality,* 2013, pp. 195-229.

[79]. P. E. Keller, Electronic noses and their applications, in *Proceedings of the IEEE Technical Applications Conference and Workshops (Northcon/95)*, 1995, pp. 116.

[80]. S. Benedetti, N. Sinelli, S. Buratti, M. Riva, Shelf life of Crescenza Cheese as Measured by Electronic Nose, *J. Dairy Sci.*, Vol. 88, Issue 9, 2005, pp. 3044–3051.

[81]. A. Bermak, S. B. Belhouari, M. Shi, D. Martinez, Pattern Recognition Techniques for Odor Discrimination in Gas Sensor Array, *Encyclopedia of Sensors*, 2005, pp. 1–17.

[82]. Z. Hai J. Wang, Detection of adulteration in camellia seed oil and sesame oil using an electronic nose, *Eur. J. Lipid Sci. Technol.*, Vol. 108, Issue 2, 2006, pp. 116–124.

[83]. J. E. Dayhoff J. M. DeLeo, Artificial neural networks, *Cancer*, Vol. 91, Issue S8, 2001, pp. 1615–1635.

[84]. M. Śliwińska, P. Wiśniewska, T. Dymerski, J. Namieśnik, W. Wardencki, Food Analysis Using Artificial Senses, *J. Agric. Food Chem.*, Vol. 62, Issue 7, 2014, pp. 1423–1448.

[85]. J. Y. Huang, X. P. Guo, Y. B. Qiu, Z. Y. Chen, Cluster and discriminant analysis of electrochemical noise data, *Electrochim. Acta*, Vol. 53, Issue 2, 2007, pp. 680–687.

[86]. Z. Haddi, Electronic nose and tongue combination for improved classification of Moroccan virgin olive oil profiles, *Food Res. Int.*, Vol. 54, 2013, pp. 1488–1498.

[87]. D. Melucci, Rapid direct analysis to discriminate geographic origin of extra virgin olive oils by flash gas chromatography electronic nose and chemometrics, *Food Chem.*, Vol. 204, 2016, pp. 263–273.

[88]. T. Praczyk, Using a probabilistic neural network and the nearest neighbour method to identify ship radiostations, *Logistyka*, Vol. 3, 2011, pp. 2235–2242.

[89]. M. Bougrini, Detection of Adulteration in Argan Oil by Using an Electronic Nose and a Voltammetric Electronic Tongue, *J. Sensors*, Vol. 2014, pp. 1–10.

[90]. A. Loutfi, S. Coradeschi, G. K. Mani, P. Shankar, J. B. Rayappan, Electronic noses for food quality: A review, *J. Food Eng.*, Vol. 144, 2015, pp. 103-111.

[91]. I. Stanimirova, B. Daszykowski, B. Walczak, Supervised methods with calibration and classification, in Chemometry in analitics, *IES*, Kraków, 2008.

[92]. Y. González Martín, Electronic nose based on metal oxide semiconductor sensors and pattern recognition techniques: characterisation of vegetable oils, *Anal. Chim. Acta*, Vol. 449, 2001, pp. 69–80.

[93]. J. Koronacki, J. Ćwik, Liniowe metody klasyfikacji, in Statystyczne systemy uczące się, 2nd ed., *Exit*, Warsaw, 2008, pp. 4–12.

[94]. M. Casale, C. Casolino, P. Oliveri, M. Forina, The potential of coupling information using three analytical techniques for identifying the geographical origin of Liguria extra virgin olive oil, *Food Chem.*, Vol. 118, 2010, pp. 163–170.

[95]. B. E. Boser, I. M. Guyon, V. N. Vapnik, A training algorithm for optimal margin classifiers, in *Proceedings of the 5th Annual Workshop on Computational Learning Theory (COLT '92)*, 1992, pp. 144–152.

[96]. N. Cristianini and B. Scholkopf, Support Vector Machines and Kernel Methods: The New Generation of Learning Machines, *AI Mag.*, Vol. 23, Issue 3, 2002, pp. 31.

[97]. N. El Barbri, E. Llobet, N. El Bari, X. Correig, B. Bouchikhi, Application of a portable electronic nose system to assess the freshness of Moroccan sardines, *Mater. Sci. Eng.*, Vol. 28, Issue 5–6, 2008, pp. 666–670.

[98]. J. Goszczyński, Klasyfikacja tekstur za pomocą SVM - Maszyny Wektorów Wspierających, *Inżynieria Rol.*, 2006, pp. 119–126.

[99]. G. Gamst, L. S. Meyers, A. J. Guarino, Analysis of variance designs: A conceptual and computational approach with SPSS and SAS, *Cambridge University Press,* 2008.

[100]. L. Marilley, S. Ampuero, T. Zesiger, M. G. Casey, Screening of aroma-producing lactic acid bacteria with an electronic nose, *Int. Dairy J.*, Vol. 14, 2004, pp. 849.

[101]. F. Winquist, E. G. Hörnsten, H. Sundgren, I. Lundström, Performance of an electronic nose for quality estimation of ground meat, *Meat Sci. Technol.*, Vol. 4, 1993, pp. 1493.

[102]. A. Jonsson, F. Winquist, J. Schnürer, H. Sundgren, I. Lundström, Electronic nose for microbial quality classification of grains, *Int. J. Food Microbiol.*, Vol. 35, Issue 2, 1997, pp. 187.

[103]. A. Spetz, F. Winquist, H. Sundgren, I. Lundström, Field effect gas sensors, in Gas Sensors, G. Sberveglieri (Ed.), *Kluwer Academic Publishers,* Dordrecht, 1992, pp. 219–280.

[104]. N. Magan, A. Pavlou, I. Chrysanthakis, Milk-sense: a volatile sensing system recognises spoilage bacteria and yeasts in milk, *Sens. Actuators B*, Vol. 72, Issue 1, 2001, pp. 28.

[105]. S. Capone, M. Epifani, F. Quaranta, P. Siciliano, A. Taurino, L. Vasanelli, Monitoring of rancidity of milk by means of an electronic nose and a dynamic PCA analysis, *Sens. Actuators B*, Vol. 78, Issue 1–3, 2001, pp. 174–179.

[106]. P. Wiśniewska, T. Dymerski, W. Wardencki, J. Namieśnik, Chemical composition analysis and authentication of whisky, *Journal of the Science of Food and Agriculture*, Vol. 95, Issue 11, 2015, pp. 2159–2166.

[107]. T. Dymerski, J. Gebicki, W. Wardencki, J. Namieśnik, Application of an electronic nose instrument to fast classification of Polish honey types, *Sensors*, Vol. 14, Issue 6, 2014, pp. 10709–10724.

[108]. W. Wardencki, T. Chmiel, T. Dymerski, P. Biernacka, Instrumental techniques used for assessment of food quality, in *Procedings of the ECOpole*, Vol. 3, Issue 2, 2009, pp. 273–279.

[109]. A. C. M. Oliveira, M. O. Balaban, Comparison of a colorimeter with a machine vision system in measuring color of Gulf of Mexico sturgeon fillets, *Appl. Eng. Agr.*, Vol. 22, 2006, pp. 583–587.

[110]. I. Camposa, Accurate Concentration Determination of Anions Nitrate, Nitrite Chloride in Minced Meat Using A Voltammetric Electronic Tongue, *Sens. Actuators B*, Vol. 149, 2010, pp. 71–78.

[111]. M. García, M. Aleixandre, M. C. Horrillo, Electronic nose for the identification of spoiled Iberian hams, in *Proceedings of the Spanish Conf. Electron Devices*, Tarragona, 2005, pp. 537–540.

[112]. N. El Barbri, An electronic nose system based on a micro-machined gas sensor array to assess the freshness of sardines, *Sens. Actuators B*, Vol. 141, Issue 2, 2009, pp. 538.

[113]. N. El Barbria, E. Llobetb, N. El Baric, X. Correigb, B. Bouchikhi, Application of a portable electronic nose system to assess the freshness of Moroccan sardines, *Mater. Sci. Eng.*, Vol. 28, 2008, pp. 666–670.

[114]. G. Olafsdottir, Prediction of microbial and sensory quality of cold smoked Atlantic salmon (Solmo salar) by electronic nose, *J. Food Sci.*, Vol. 70, Issue 9, 2005, pp. 563–574.

[115]. J. Laothawornkitkul, Discrimination of plant volatile signatures by an electronic nose: a potential technology for plant pest disease monitoring, *Env. Sci Technol.*, Vol. 42, 2008, pp. 8433–8439.

[116]. C. Di Natale, G. Olafsdottir, S. Einarsson, E. Martinelli, R. Paolesse, A. D'Amico, Comparison and integration of different electronic noses for freshness evaluation of cold-fish fillets, *Sens. Actuators B*, Vol. 77, 2001, pp. 572–578.

[117]. F. Korel, D. A. Luzuriaga, M. Ö. Balaban, Objective quality assessment of raw Tilapia (Oreochromis niloticus) fillets using electronic nose and machine vision, *J. Food Sci.*, Vol. 66, 2001, pp. 1018–1024.

[118]. X. Hu, P. K. Mallikarjunan, D. Vaughan, Development of non-destructive methods to evaluate oyster quality by electronic nose technology, *Sens. Instrumen. Food Qual.*, Vol. 2, Issue 1, 2008, pp. 51.

[119]. A. Legin, A. Rudnitskaya, B. Seleznev, G. Sparfel, C. Doré, Electronic tongue distinguishes onions and shallots, in *Procedings of the ISHS Acta Hortic. 634 XXVI Int. Hortic. Congr. IV Int. Symp. Taxon. Cultiv. Plants.*, Toronto, 2004.

[120]. K. Beullens, P. Mészáros, S. Vermeir, D. Kirsanov, A. Legin, S. Buysens, N. Cap, B. M. Nicolaï, J. Lammertyn, Analysis of tomato taste using two types of electronic tongues, *Sens. Actuators B*, Vol. 131, 2008, pp. 10–17.

[121]. A. Rudnitskaya, Analysis of apples varieties – comparison of electronic tongue with different analytical techniques, *Sens. Actuators B*, Vol. 116, 2006, pp. 23–28.

[122]. P. Pani, A. A. Leva, M. Riva, A. Maestrelli, D. Torreggiani, Influence of an osmotic pre-treatment on structure-property relationships of air-dehydrated tomato slices, *J. Food Eng.*, Vol. 86, Issue 1, 2008, p. 105.

[123]. M. Santonico, A. Bellincontro, D. De Santis, C. Di Natale, F. Mencarelli, Electronic nose to study postharvest dehydration of wine grapes, *Food Chem.*, Vol. 121, 2010, pp. 789–796.

[124]. C. Oliveros, R. Boggia, M. Casale, C. Armanino, M. Forina, Optimisation of a new headspace mass spectrometry instrument, *J. Chromatogr. A*, Vol. 1076, 2005, pp. 7–15.

[125]. G. Pioggia, M. Ferro, F. Di Francesco, Towards a Real-Time Transduction and Classification of Chemoresistive Sensor Array Signals, *IEEE Sens. J.*, Vol. 7, Issue 2, 2007, pp. 237-243.

[126]. R. Aparicio, R. Aparicio-Ruíz, Authentication of vegetable oils by chromatographic techniques, *J. Chromatogr. A*, Vol. 881, Issue 1–2, 2000, pp. 93–104.

[127]. X. Zheng, Y. Lan, J. Zhu, J. Westbrook, W. C. Hoffmann, R. E. Lacey, Rapid Identification of Rice Samples Using an Electronic Nose, *J. Bionic Eng.*, Vol. 6, 2009, pp. 290–297.

[128]. T. Borjesson, Detection of off-odours in grains using an electronic nose, in *Proceedings of the Olfaction Electron. Nose: 3rd Int. Symp.*, Toulouse, 1996.

[129]. T. Börjesson, T. Eklöv, A. Jonsson, H. Sundgren, J. Schnürer, Electronic Nose for Odor Classification of Grains, *Cereal Chem.*, Vol. 73, 1996, pp. 457–461.

[130]. S. Panigrahi, M. K. Misra, S. Willson, Evaluations of fractal geometry and invariant moments for shape classification of corn germplasm, *Comput. Electron. Agric.*, Vol. 20, Issue 1, 1998, pp. 1–20.

[131]. J. W. Gardner, H. V. Shurmer, T. T. Tan, Application of an electronic nose to the discrimination of coffees, *Sens. Actuators B*, Vol. 6, Issue 1–3, 1992, pp. 71.

[132]. E. Schaller, J. Bosset, F. Escher, 'Electronic Noses' and Their Application to Food, *LWT Food Sci. Technol.*, Vol. 31, Issue 4, 1998, pp. 305–316.

[133]. A. Legin, A. Rudnitskaya, L. Lvova, Y. Vlasov, C. Di Natale, A. D'Amico, Evaluation of Italian wine by the electronic tongue: recognition, quantitative analysis and correlation with human sensory perception, *Anal. Chim. Acta*, Vol. 484, Issue 1, 2003, pp. 33–44.

[134]. M. Falasconi, M. Pardo, G. Sberveglieri, I. Riccò, A. Bresciani, The novel EOS835 electronic nose and data analysis for evaluating Coffee ripening, *Sens. Actuators B*, Vol. 110, 2005, pp. 73–80.

[135]. H. Yu, J. Wang, Discrimination of LongJing green-tea grade by electronic nose, *Sens. Actuators B*, Vol. 122, 2007, p. 134.

[136]. M. Palit, Classification of Black Tea Taste and Correlation With Tea Taster's Mark Using Voltammetric Electronic Tongue, *IEEE Trans. Instrum. Meas.*, Vol. 59, 2010, pp. 2230–2239.

[137]. Z. Yanga, Identification of coumarin-enriched Japanese green teas and their particular flavor using electronic nose, *J. Food Eng.*, Vol. 92, 2009, pp. 312–316.

[138]. A. Ghosh, P. Tamuly, N. Bhattacharyya, B. Tudu, R. Gogoi, N. Bandyopadhyay, Estimation of theaflavin content in black tea using electronic tongue, *J. Food Eng.*, Vol. 110, 2012, pp. 71–79.

[139]. C. Zhang, K. S. Suslick, Colorimetric sensor array for soft drink analysis, *J. Agric. Food Chem.*, Vol. 55, Issue 2, 2007, pp. 237–255.

[140]. P. Ciosek, Z. Brzózka, W. Wróblewski, Classification of beverages using a reduced sensor array, *Sens. Actuators B*, Vol. 103, Issue 1–2, 2004, pp. 76-80.

[141]. V. Martina, Development of an electronic tongue based on a PEDOT-modified voltammetric sensor, *Analytical and Bioanalytical Chemistry,* 387, 6, 2007, pp. 2011–2110.

[142]. S. Montag, M. Frank, H. Ulmer, D. Wernet, W. Göpel, H. G. Rammensee, 'Electronic nose' detects major histocompatibility complex-dependent prerenal and postrenal odor components, *Immunology*, Vol. 98, Issue 16, 2001, pp. 9249–9254.

[143]. C. Pinheiro, C. M. Rodrigues, T. Schäfer, J. G. Crespo, Monitoring the Aroma Production During Wine-Must Fermentation with an Electronic Nose, *Biotechnol. Bioeng.*, Vol. 77, Issue 6, 2002, p. 632.

[144]. P. Ragazzo-Sanchez, J. A. Chalier, D. Chevalier, C. Calderon-Santoyo, M. Ghommidh, Identification of Different Alcoholic Beverages by Electronic Nose Coupled to GC, *Sens. Actuators B*, Vol. 134, 2008, pp. 43–48.

[145]. M. Liu, Application of electronic nose in Chinese spirits quality control and flavour assessment, *Food Control*, Vol. 26, Issue 2, 2012, pp. 564–570.

[146]. J. A. Ragazzo-Sanchez, P. Chalier, D. Chevalier, M. Calderon-Santoyo, C. Ghommidh, Identification of different alcoholic beverages by

electronic nose coupled to GC, *Sens. Actuators B*, Vol. 134, Issue 1, 2008, pp. 43-51.

[147]. H. H. Wang, Assessment of solid state fermentation by a bioelectronic artificial nose, *Biotech. Advan.*, Vol. 11, 1993, pp. 701–710.

[148]. M. A. Drake, P. Gerard, J., Kleinhenz, W. Harper, Application of an electronic nose to correlate with descriptive sensory analysis of aged Cheddar cheese, *LWT - Food Sci. Technol.*, Vol. 36, Issue 1, 2003, pp. 13–20.

[149]. N. Bhattacharya, B. Tudu, A. Jana, D. Ghosh, R. Bandhopadhyaya, M. Bhuyan, Preemptive identification of optimum fermentation time for black tea using electronic nose, *Sens. Actuators B*, Vol. 131, Issue 1, 2008, p. 110.

[150]. J. E. Haugen, Rapid control of smoked Atlantic salmon (Salmo salar) quality by electronic nose: Correlation with classical evaluation methods, *Sens. Actuators B*, Vol. 116, Issue 1–2, 2006, pp. 72.

[151]. R. Aparicio, S. M. Rocha, I. Delgadillo, M. T. Morales, Detection of Rancid Defect in Virgin Olive Oil by the Electronic Nose, *J. Agric. Food Chem.*, Vol. 48, 2000, pp. 853–860.

[152]. J. Brezmes, E. Llobet, X. Vilanova, G. Saiz, X. Correig, Fruit ripeness monitoring using an Electronic Nose, *Sens. Actuators B*, Vol. 69, Issue 3, 2000, pp. 223–229.

[153]. S. Saevels, J. Lammertyn, A. Z. Berna, E. A. Veraverbeke, C. Di Natale, B. M. Nicolaï, An electronic nose and a mass spectrometry-based electronic nose for assessing apple quality during shelf life, *Postharvest Biol. Technol.*, Vol. 31, Issue 1, 2004, pp. 9–16.

[154]. U. Herrmann, Monitoring apple flavor by use of quartz microbalances, *Anal. Bioanal. Chem.*, Vol. 372, Issue 5–6, 2002, pp. 611–619.

[155]. S. Capone, P. Siciliano, F. Quaranta, R. Rella, M. Epifani, L. Vasanelli, Analysis of vapours and foods by means of an electronic nose based on a sol-gel metal oxide sensors array, *Sens. Actuators B*, Vol. 69, Issue 3, 2000, p. 230.

[156]. M. Penza, G. Cassano, Recognition of adulteration of Italian wines by thin-film multisensor array and artificial neural networks, *Anal. Chim. Acta*, Vol. 509, Issue 2, 2004, pp. 159–177.

[157]. H. Yu, J. Wang, H. Zhang, Y. Yu, C. Yao, Identification of green tea grade using different feature of response signal from E-nose sensors, *Sens. Actuators B*, Vol. 128, Issue 2, 2008, pp. 455–461.

[158]. S. López-Feria, S. Cárdenas, J. A. García-Mesa, M. Valcárcel, Simple and rapid instrumental characterization of sensory attributes of virgin olive oil based on the direct coupling headspace-mass spectrometry, *J. Chromatogr. A*, Vol. 1188, Issue 2, 2008, pp. 308-313.

[159]. C. Apetrei, M. Ghasemi-Varnamkhasti, I. Mirela Apetrei, Olive Oil and Combined Electronic Nose and Tongue, in *Electronic Noses and Tongues in Food Science*, 2016, pp. 277–289.

[160]. C. Di Natale, An electronic nose for food analisys, *Sens. Actuators B*, Vol. 44, Issue 1–3, 1997, pp. 521-527.

[161]. C. Di Natale, A. Macagnano, E. Martinelli, R. Paolesse, E. Proietti, A. D'Amico, The evaluation of quality of post-harvest oranges and apples by means of an electronic nose, *Sens. Actuators B*, Vol. 78, Issue 1–3, 2001, pp. 26–32.

[162]. C. Di Natale, Electronic nose based investigation of the sensorial properties of peaches and nectarines, *Sens. Actuators B*, Vol. 77, Issue 1–2, 2001, p. 561–570.

[163]. N. García-Villar, S. Hernández-Cassou, J. Saurina, Characterization of wines through the biogenic amine contents using chromatographic techniques and chemometric data analysis, *J. Agric. Food Chem.*, Vol. 55, Issue 18, 2007, pp. 7453–7461.

[164]. M. J. Cejudo-Bastante, I. Hermosín-Gutiérrez, M. S. Pérez-Coello, Micro-oxygenation and oak chip treatments of red wines: Effects on colour-related phenolics, volatile composition and sensory characteristics. Part II: Merlot wines, *Food Chem.*, Vol. 124, Issue 3, 2011, pp. 738–748.

[165]. L. Gil, J. M. Barat, I. Escriche, E. Garcia-Breijo, R. Martínez-Máñez, and J. Soto, An electronic tongue for fish freshness analysis using a thick-film array of electrodes, *Microchim. Acta*, Vol. 163, Issue 1–2, 2008, pp. 121–129.

[166]. M. L. Rodríguez-Méndez, M. Gay, C. Apetrei, J. A. de Saja, Biogenic amines and fish freshness assessment using a multisensor system based on voltammetric electrodes. Comparison between CPE and screen-printed electrodes, *Electrochim. Acta*, Vol. 54, 2009, pp. 7033–7041.

[167]. T. U. Tran, K. Suzuki, H. Okadome, S. Homma, K. Ohtsubo, Analysis of the tastes of brown rice and milled rice with different milling yields using a taste sensing system, *Food Chem.*, Vol. 88, 2004, pp. 557–566.

[168]. Z. Wei, J. Wang, X. Zhang, Monitoring of quality and storage time of unsealed pasteurized milk by voltammetric electronic tongue, *Electrochim. Acta*, Vol. 88, 2013, pp. 231–239.

[169]. I. Escriche, M. Kadar, E. Domenech, L. Gil-Sánchez, A potentiometric electronic tongue for the discrimination of honey according to the botanical origin. Comparison with traditional methodologies: Physicochemical parameters and volatile profile, *J. Food Eng.*, Vol. 109, 2012, pp. 449–456.

[170]. Z. Wei, J. Wang, W. Liao, Technique potential for classification of honey by electronic tongue, *J. Food Eng.*, Vol. 94, Issue 3–4, 2009, pp. 260–266.

[171]. N. Major, K. Marković, M. Krpan, G. Sarić, M. Hruškar, N. Vahčić, Rapid honey characterization and botanical classification by an electronic tongue, *Talanta*, Vol. 85, Issue 1, 2011, pp. 569–574.

[172]. A. Legin, A. Rudnitskaya, B. Seleznev, Y. Vlasov, Electronic tongue for quality assessment of ethanol, vodka and eau-de-vie, *Anal. Chim. Acta*, Vol. 534, Issue 1, 2005, pp. 129–135.

[173]. L. Lvova, R. Paolesse, C. Di Natale, A. D'Amico, Detection of alcohols in beverages: An application of porphyrin-based Electronic tongue, *Sensors Actuators, B Chem.*, Vol. 118, Issue 1–2, 2006, pp. 439–447.

[174]. W. Novakowski, M. Bertotti, T. R. L. C. Paixão, Use of copper and gold electrodes as sensitive elements for fabrication of an electronic tongue: Discrimination of wines and whiskies, *Microchem. J.*, Vol. 99, Issue 1, 2011, pp. 145–151.

[175]. M. L. Rodríguez-Méndez, C. Apetrei, J. A. de Saja, Evaluation of the polyphenolic content of extra virgin olive oils using an array of voltammetric sensors, *Electrochim. Acta*, Vol. 53, 2008, pp. 5867–5872.

[176]. C. Krantz-Rülcker, M. Stenberg, F. Winquist, I. Lundström, Electronic tongues for environmental monitoring based on sensor arrays and pattern recognition: a review, *Anal. Chim. Acta*, Vol. 426, Issue 2, 2001, pp. 217–226.

[177]. L. Capelli, S. Sironi, R. Del Rosso, Electronic Noses for Environmental Monitoring Applications, *Sensors*, Vol. 14, Issue 11, 2014, pp. 19979–20007.

[178]. D. Hodgins, The development of an electronic 'nose' for industrial and environmental applications, *Sensors Actuators B Chem.*, Vol. 27, Issue 1–3, 1995, pp. 255–258.

[179]. M. Abbas, G. Moustafa, W. Gopel, Multicomponent analysis of some environmentally important gases using semiconductor tin oxide sensors, *Anal. Chim. Acta*, Vol. 431, Issue 2, 2001, pp. 181–194.

[180]. O. Helli, M. Siadat, M. Lumbreras, Qualitative and quantitative identification of H2S/NO2 gaseous components in different reference atmospheres using a metal oxide sensor array, *Sensors Actuators B Chem.*, Vol. 103, Issue 1, 2004, pp. 403–408.

[181]. E. J. Wolfrum, R. M. Meglen, D. Peterson, J. Sluiter, Metal oxide sensor arrays for the detection, differentiation, and quantification of volatile organic compounds at sub-parts-per-million concentration levels, *Sensors Actuators B Chem.*, Vol. 115, Issue 1, 2006, pp. 322–329.

[182]. T. Dymerski, J. Gębicki, P. Wiśniewska, M. Sliwińska, W. Wardencki, J. Namieśnik, Application of the electronic nose technique to differentiation between model mixtures with COPD markers, *Sensors*, Vol. 13, Issue 4, 2013, pp. 5008–5027.

[183]. D. S. Lee, J. K. Jung, J. W. Lim, J. S. Huh, D. D. Lee, Recognition of volatile organic compounds using SnO2 sensor array and pattern recognition analysis, *Sensors Actuators B Chem.*, Vol. 77, Issue 1, 2001, pp. 228–236.

[184]. A. J. Lilienthal, A. Loutfi, T. Duckett, Airborne Chemical Sensing with Mobile Robots, *Sensors*, Vol. 6, Issue 11, 2006, pp. 1616–1678.

[185]. CEN. EN 13725:2003. Air Quality—Determination of Odour Concentration by Dynamic Olfactometry, *Comité Européen de Normalisation,* Brussels, Belgium, 2007.

[186]. T. H. Misselbrook, P. J. Hobbs, K. C. Persaud, Use of an Electronic Nose to Measure Odour Concentration Following Application of Cattle Slurry to Grassland, *J. Agric. Eng. Res.*, Vol. 66, Issue 3, 1997, pp. 213–220.

[187]. J. H. Sohn, N. Hudson, E. Gallagher, M. Dunlop, L. Zeller, M. Atzeni, Implementation of an electronic nose for continuous odour monitoring in a poultry shed, *Sensors Actuators B Chem.*, Vol. 133, Issue 1, 2008, pp. 60–69.

[188]. S. Sironi, L. Capelli, P. Céntola, R. Del Rosso, M. Grande, Continuous monitoring of odours from a composting plant using electronic noses, *Waste Manag.*, Vol. 27, Issue 3, 2007, pp. 389–397.

[189]. R. M. Stuetz, R. A. Fenner, G. Engin, Characterisation of wastewater using an electronic nose, *Water Res.*, Vol. 33, Issue 2, 1999, pp. 442–452.

[190]. R. M. Stuetz, S. George, R. A. Fenner, S. J. Hall, Monitoring wastewater BOD using a non-specific sensor array, *J. Chem. Technol. Biotechnol.*, Vol. 74, Issue 11, 1999, pp. 1069–1074.

[191]. R. Baby, M. Cabezas, E. W. de Reca, Electronic nose: a useful tool for monitoring environmental contamination, *Sensors Actuators B Chem.*, Vol. 69, Issue 3, 2000, pp. 214–218.

[192]. A. Lamagna, S. Reich, D. Rodríguez, A. Boselli, D. Cicerone, The use of an electronic nose to characterize emissions from a highly polluted river, *Sensors Actuators B Chem.*, Vol. 131, Issue 1, 2008, pp. 121–124.

[193]. A. Catarina Bastos N. Magan, Potential of an electronic nose for the early detection and differentiation between Streptomyces in potable water, *Sensors Actuators B Chem.*, Vol. 116, Issue 1, 2006, pp. 151–155.

[194]. J. Nicolas, A. C. Romain, V. Wiertz, J. Maternova, P. Andre, Using the classification model of an electronic nose to assign unknown malodours to environmental sources and to monitor them continuously, *Sens. Actuators B*, Vol. 69, 2000, pp. 366–371.

[195]. S. Zampolli, An electronic nose based on solid state sensor arrays for low-cost indoor air quality monitoring applications, *Sens. Actuators B*, Vol. 101, Issue 1–2, 2004, pp. 39–46.

[196]. L. Capelli, S. Sironi, P. Céntola, R. Del Rosso, M. Grande, Electronic noses for the continuous monitoring of odours from a wastewater treatment plant at specific receptors: Focus on training methods, *Sens. Actuators B*, Vol. 131, Issue 1, 2008, pp. 53–62.

[197]. F. D. Francesco, B. Lazzerini, F. Marcelloni, G. Pioggia, An electronic nose for odour annoyance assessment, *Atmos. Environ.*, Vol. 35, 2001, pp. 1225–1231.

[198]. S. Fuchs, P. Strobel, M. Siadat, M. Lumbreras, Evaluation of unpleasant odor with a portable electronic nose, *Mater. Sci. Eng. C*, Vol. 28, Issue 5–6, 2008, pp. 949–953.

[199]. D. Fernandes, M. T. Gomes, Development of an electronic nose to identify and quantify volatile hazardous compounds, *Talanta*, Vol. 77, Issue 1, 2008, pp. 77–83.

[200]. L. Zhang, Chaos based neural network optimization for concentration estimation of indoor air contaminants by an electronic nose, *Sens. Actuators B*, Vol. 189, 2013, pp. 161–167.

[201]. L. Zhang, On-line sensor calibration transfer among electronic nose instruments for monitoring volatile organic chemicals in indoor air quality, *Sensors Actuators, B Chem.*, Vol. 160, Issue 1, 2011, pp. 899–909.

Index

www.ingramcontent.com/pod-product-compliance
Lightning Source LLC
Chambersburg PA
CBHW050459190326
41458CB00005B/1355